Contents

SPACE GRAVITY & DARK ENERGY

A journey beyond Newton and Einstein

DR. ALAGAR RAMANUJAM

&

VIJAY ARORA

ISBN 979-8-89233-566-9

Preface

This book explores the unsolved challenges and mysteries in physics remaining for the past few centuries. Essentially this book is a follow-up to the work of Newton and Einstein in the field of Gravity. We have been working for twenty years to find broader solutions for problems related to gravity and the anti-gravity dark energy.

Gravity is a term both familiar and mysterious to many of us. Though we have a formula to measure its magnitude between any two objects, scientists don't fully understand how objects provide push or pull on each other kept at a faraway distance. Regarding its cause, great giants in the field of science have different views and sometimes even opposite views. According to Jayanth Narlikar, "Gravity remains shrouded in mystery and Nature has more secrets up her sleeve".

Newtonian mechanics states that every object in the universe exerts a force on every other object. The magnitude of this force is determined by Newton's law of gravity. However, Einstein disagrees with this idea and claims that an object does not exert a force on another object. According to him, an object acts on the space surrounding it and curves it. In curved space, objects move along curved paths without a force acting on them. This creates the effect of gravity. Einstein also disagrees with Newton's idea of "action at a distance" because it suggests infinite velocities.

Apart from holding different views about gravity, both the Newtonian and Einsteinian concepts of gravity have serious limitations. While Newton deduced his law of gravity from Kepler's laws of planetary motion, he could not derive it by framing suitable axioms. When pressed, he kept reiterating that he did not want to frame any axiom about the source of gravity. When two objects move towards each other, Newton argued that one may pull the other or the space around them may push them together. I don't know which is true.

Einstein used Galileo's experimental results to develop his theory of gravity. Galileo had demonstrated that two objects of different masses dropped from the same height reached the ground at the same time. When someone reminded him that Galileo's result had no theoretical proof, Einstein said Galileo's demonstration was enough, even without a theoretical proof.

In both Newton and Einstein's formalism on gravity, the term "mass of an object" remains undefined. To truly comprehend gravity, we need to understand the mass of an object because the formula for gravitational force includes the mass factor. Furthermore, neither Newton gives a mechanism as to how a mass affects another mass far away nor Einstein explains how a mass can warp limitless space.

Both Newton and Einstein gave a physical reality to matter and never considered it as born out of a source. Recently, some scientists, like John Wheeler, have speculated about the source of matter. Wheeler believes that a particle could be seen as an excitation of space.

The above points show that we still don't know where matter comes from or what causes gravity and anti-gravity dark energy. Scientists are studying if a new theory of gravity is needed. However, they don't know yet if they need a completely new theory or just changes to the current one. This has motivated us to develop a new theory of gravity. It aims to solve some of the limitations of Newtonian and Einsteinian gravity.

In the first two chapters, we talk about Newtonian and Einsteinian ideas of gravity. Then, in the third chapter, we talk about the newer Vethathirian concept of gravity. The Vethathirian theory of gravity starts with two basic ideas in the form of axioms which lead to many other significant results.

One important aspect of this new theory is that Space is not seen as empty by Vethathirian gravity. Instead, it is seen as a "substance" of Consciousness and Energy. Through a self-transformation process, it becomes the universe.

Instead of using the phrase "Let m be the mass of a body," the Vethathirian concept starts by explaining what mass is. It also provides a formula for the mass of a body based on its causative factors. The Vethathirian theory is a significant scientific development because it defines mass. The mass remains undefined in both Newtonian and Einsteinian theories of gravity.

Vethathirian theory uses mass formula to derive Newton's law of gravity. It doesn't rely on Kepler's laws. This derivation is the first of its kind. It also automatically provides a proof for Galileo's experimental result. The result is that two different masses, dropped from the same height above the ground, will reach the ground at the same time. Such a proof has been pending for the last few centuries.

In chapter III, we review the popular statement that one particle attracts another particle because Newton never made that statement. Even in Newton's time, he had many questions about his law of gravity that bothered him. These questions are discussed here with appropriate explanations.

Both Newton and Einstein's theories of gravity apply to the entire solar system and explain how galaxies work. However, when it comes to discussing the behavior of the universe as a whole, they fail. Both concepts expect the universe to slow down due to gravity, so the accelerated expansion of the universe is surprising and unexplainable for them. To resolve this enigma, an anti-gravity repulsive force has been introduced for now. The source of the antigravity force is completely unknown. So, it is called Dark Energy. The study of dark energy is now a significant area of research in cosmology.

We have now come to see another application of our mass formula. The Schwarzschild metric is a solution to Einstein's field equations for gravity. It describes the curvature of spacetime around a non-rotating spherical mass. In the third chapter, we review how the Schwarzschild metric is derived. We also discuss that in conjunction with the two axioms of Vethathirian gravity. The expression for ds^2, which is the square of the spacetime interval, is obtained in terms of the fundamental space parameters in Vethathirian gravity. We consider this as a significant result emerging from the axioms of Vethathirian gravity.

The mass formula derived in this chapter is the first ever equation that connects mass and space in physics. We are pleased to see that our mass formula includes terms for both compressive gravity (C) and the antigravity Dark Energy (R). The mass formula helps us to derive both the Newton's law of gravity and the Schwarzschild metric relation from a different perspective.

A word of comfort and caution. A scholar working in scientific fields like molecular spectroscopy, nuclear magnetic resonance, artificial intelligence, or rocket launching, can afford to skip the third chapter. They lose nothing if they completely omit the third chapter. Their life and work will still be normal. This is the comfort. However, if someone wants to work in the field of cosmology and wants to understand the universe, he can't afford to skip the third chapter. The discussion given in the third chapter will be of immense help to him. This is the caution. Einstein stands as a witness for this.

Einstein's lack of understanding about the issues discussed in the third chapter led to his first cosmology paper being a failure. He never wrote a second cosmology paper. He never defined or explained what mass is. When he formulated a new theory on gravity, he didn't mention what causes it. He believed in the popular idea that mass causes gravity which Newton himself never accepted. Furthermore, Einstein didn't seek theoretical proof for Galileo's experiment which was the basis of his general theory of relativity. He wrote important papers as a patent office clerk, but as a top man in science, he didn't write any significant papers in cosmology.

The Big Bang theory is facing problems because it hasn't addressed important issues like the cause of gravity, anti-gravity, and the definition of mass. One of its latest problems is not being able to explain the acceleration of the expansion of the universe. We believe that to do serious work in cosmology, we must understand fundamental issues that have remained unresolved for centuries. As this book provides serious discussions on long standing fundamental issues, this book may be considered as a reliable gateway to cosmology.

The Vethathirian theory of gravity which begins from the Beginning, that is the Space, is a great contribution to science. Due to its beginning from the Beginning, it has brought many significant results such as the mass formula, derivation of the law of gravity, etc. which the Newtonian and Einsteinian concepts could not have imagined at all.

We remain extremely thankful to Arulnithi Uma Vethathiri who has been and is a source of inspiration for us in developing the Vethathirian theory of gravity presented in this book. Our sincere thanks are due to Arulnithi Mr. R. Prakash and Arulnithi Mrs. D. Padma Priya for very many useful discussions.

Our sincere thanks also to Mrs. Thayar Alagar Ramanujam and Mrs. Anjali Arora for their consistent support and encouragement in bringing out this book.

Our profound thanks are due to Notion Press, CHENNAI for their sincere efforts in bringing out this book in the best possible manner.

– Authors'

Chapter 1

Newtonian Concept of Gravity

Gravity is the weakest of the fundamental forces operating in the universe and at the same time, is the most problematic. The cause of gravity still remains unsettled. Science understands the causes of forces like electromagnetic and nuclear force, but not gravity. Newton made an interesting remark about gravity and its effects on the heavens and sea. He said that we have explained these phenomena but haven't identified the cause of gravity. (Book III Newton's Principia)

Newton who introduced gravity to the world, never said that every particle attracts every other particle. People who came after him made such a statement in the name of Newton much against his wish. In 1692, Newton's close friend Richard Bentley wrote in a textbook that every material particle attracts each other. When Newton found out, he wrote to Richard Bentley - "You sometimes speak of gravity as essential and inherent to matter. Pray do not ascribe that notion to me, for, the cause of gravity is what I do not pretend to know and would therefore take more time to consider of it".

If two particles, m_1 and m_2, move towards each other, Newton visualizes two possible causes. One cause could be m_1 (or m_2) exerting a force on m_2 (or m_1) to pull it. Another cause could be the space around the particles pushing them towards each other. When questioned, which of the above two possibilities is happening, Newton said, "I don't know...." In the former case, the cause is material and in the latter case, it is non-material. Newton said he wasn't sure if gravity is due to particles or space, but textbooks misquote him.

Let us sketch Newton's life which could be a motivator to many youngsters in schools and colleges. Isaac Newton was born in England on a Christmas Day, 25th December 1642, his father having died three months earlier. Even though Newton's grandfather was rich from ownership in land and livestock, his father was illiterate and a farmer in a village. When Newton was still a baby, his mother married again to a sixty-three-year-old rector named Barnabas Smith. Barnabas refused to let Newton live with them. As a result, he ended up staying with his maternal grandmother.

Newton started to develop hatred towards his mother and step father who had abandoned him. As a child, he also had this idea of threatening his parents and burning them and their house. It is not surprising Newton grew into an embittered, isolated, and sometimes a cruel man. Newton was also argumentative and solitary and remained unmarried until his death. Despite this, he is considered one of the greatest scientific geniuses.

Newton was sent to a school at Grantham, a town close to his native village, Woolsthorpe. Because he wasn't good at studying, his mother took him out of school. She then asked him to manage her estate, but he wasn't very interested in that either. Later, when Newton himself expressed his willingness to continue his studies, his mother was reluctant to send him back to school. However, his uncle William Ayscough was supportive of Newton's wish and persuaded his mother to send him back to school. Newton returned to the Free Grammar School in Grantham in 1660 and completed his school education.

Newton wasn't great in school, but in 1661 he got the chance to go to Trinity College. He planned to study Law, but soon became interested in Mathematics. In 1665, he got his bachelor's degree in mathematics and philosophy. But the University closed in 1665-1666 due to the Great Plague, so Newton went back to his family home in Woolsthorpe, Lincolnshire. This break from his studies at Cambridge turned out to be a significant period in his life. During this time, Newton made several discoveries and advancements in science and mathematics. He created differential calculus and integral calculus, which are now fundamental in math and physics.

He came up with the three laws of motion, which are the foundation of classical mechanics. He also laid the groundwork for the theory of gravitation and made important discoveries about light. The year Newton spent away from Cambridge due to the plague outbreak turned out to be a very productive period for Newton. This period in the history of physics is like Einstein's miracle year of 1905 when he published three earth shaking papers. When Newton was asked how he made amazing discoveries in light and calculus, he answered, "By thinking about them."

When Cambridge reopened, Newton resumed his position as a student and eventually became a fellow of Trinity College. He continued to work on his ideas and further refined them. In academia, his work on calculus, laws of motion, and universal gravitation came to be appreciated and acknowledged.

Introducing Gravity

Of the fundamental forces operating in the universe, gravity is the weakest. However, because of its presence everywhere, it is also the mightiest. It is considered an attractive force existing between any two objects of the universe. While gravity allows grass to defy it and to grow up against it, at the level of the solar system it whips up the earth to go around the sun. In the former case, gravity looks weak but in the latter case, gravity looks all-powerful.

Before 600 B.C., an Indian sage named Kanada Maharishi talked about Gurutwa, the force of gravity. This is discussed in his Sanskrit book called Vaisheshika Darshanam. Later, Newton independently came to recognize this force when he observed a falling apple in 1666. This event is often associated with his realization about the force of gravity. It inspired Newton to explore the concept of gravity that eventually led to his groundbreaking work on the subject.

It took over ten years for Newton to develop his formula to measure the force of gravity between objects. Around the same time period, many scholars such as Wren, Robert Hooke, and Halley were also engaged in finding a law for gravity. However, Newton succeeded in deducing his law of gravity before anybody else, by using Kepler's laws of planetary motion.

Kepler's Laws of Planetary Motion

Kepler brought out his laws of planetary motions from the experimental data observed and laws recorded by Tycho Brahe. Newton used these to develop his law of gravitation.

Johannes Kepler was a prominent German astronomer and mathematician who lived from 1571 to 1630. A protestant by birth, he was a provincial school teacher of humble origin, known only to a few mathematicians. Kepler was sent to a protestant seminary school in the provincial town of Maulbronn to be educated for the clergy. However, instead of becoming a clergy, Kepler obtained a teaching job at the Protestant school in Graz, Austria in 1594. Even though Kepler was a brilliant thinker and a lucid writer, he could not excel as a teacher and that troubled him. During that time in Austria, Archduke Ferdinand, who was a strong Catholic, would punish anyone who didn't practice the Roman Catholic faith. The punishment could be a fine of one tenth of their income or even death. Kepler disagreed with the Catholic beliefs in Graz, causing tension and discomfort in his position. In addition, prayer, books, and hymns deemed heretical were forbidden. This coupled with financial difficulties forced Kepler's school to close.

Kepler chose to self-exile himself. To quote him "Hypocrisy I have never learned. I am in earnest about faith; I do not play with it". Kepler left Graz in 1600 with his wife and stepdaughter to seek new opportunities in Prague. He left hoping to meet Tycho Brahe, a Danish nobleman and astronomer well-known in observational astronomy. Kepler was seeking access to the extensive observational data that Tycho Brahe had gathered over the years.

Kepler's decision to leave Graz for Prague to meet Tycho Brahe was a personal decision but is significant in the history of physics. His decision led to a meeting between the two and Tycho took Kepler as his student. They analyzed and interpreted observational data together. At first, Tycho and Kepler didn't get along because they were very different. Tycho was rich and powerful, while Kepler was poor and ordinary. However, Kepler earned Tycho's trust through hard work and dedication. This partnership set the stage for Kepler's future achievements.

Tycho Brahe was a great thinker of his time. He built a fantastic observatory on Hven Island, near Copenhagen. It had modern instruments for studying the stars. He had a habit of excessive drinking, especially at royal dinners and functions. Because of this habit, Brahe gradually lost his health and died suddenly in 1601. After Brahe died, his data and instruments were passed down to Kepler, and they played a key role in his development of the laws of planetary motion.

In Kepler's time, people believed in the geocentric model of the universe. This model said that Earth was the center and other celestial bodies orbited it. Kepler was influenced by Copernicus's heliocentric model. This model put the sun in the center of the solar system and the planets orbit around it. Kepler wanted to permanently settle the conflict between the two concepts. Kepler spent eight years meticulously studying Tycho's observations and found the data supported the heliocentric model of the universe.

However, Kepler faced several challenges in reconciling his support for the heliocentric model. He had a strong belief that planets were divine bodies that should move only in circles. Kepler expected Tycho's observations to support his belief that planets move along circles. However, the observations pointed to elliptical orbits. He found it hard to accept that the divine planets would have an asymmetric motion that would deviate from perfectness of a circle. Kepler's belief was so strong that he started to doubt the veracity and the correctness of Tycho's observations. He verified and reverified the readings of Tycho, but all the data pointed to elliptical orbits of planets.

After a lot of mental struggles, Kepler finally put aside his personal belief and accepted Tycho's data. He pacified his mind saying that Earth itself is not perfect. It has aspects that are good, bad, and ugly like love, hatred, war, famine etcetera. Earth is after all a collection of stones and mud. He concluded that all other planets must be like earth. That being the case, their orbits could be anything, perfect circles, or imperfect circles (ellipses).

Once Kepler came out of his "blind belief" he became more open and receptive. More clarity emerged from the study of the observations of Tycho Brahe. That the orbits of planets were not perfect circles, but ellipses with sun at one of the foci was a significant breakthrough. This breakthrough led to the formulation of Kepler's laws of planetary motion that support the heliocentric model.

1. Every planet moves in an elliptical orbit with the Sun at one of the foci.

2. The radius vector linking the Sun and the planet sweeps equal areas in equal intervals of time. In other words, a planet moves faster when it is closer to the sun and slower when it is farther away. This law describes the unequal speed of planets along the elliptical orbits.

3. The square of the orbital period (time taken for a planet to complete one orbit around the sun) of the planet is directly proportional to the cube of semi-major axes of the orbit. This law implies that the period of a planet to orbit the Sun increase rapidly with the radius of the orbit. This law, also called the Harmonic Law, mathematically relates distance and time in planetary orbits.

Kepler's work played a crucial role in transitioning from the geocentric to the heliocentric model. This transition greatly influenced our understanding of planetary motion.

Thirty-four years after the death of Kepler, Thomas Street, an astronomer, and a close friend of Newton, introduced kepler's work to Newton. This had a profound impact on Newton, and he was able to combine his mechanics with Kepler's Laws of planetary motion. With this, Newton was able to deduce a formula for the magnitude of the force of gravity between the sun and a planet going around the sun. We give below Newton's original deduction of his law of gravity.

Newton's Deduction of His Gravity Law

Consider a planet of mass m revolving around the sun of mass M in a nearly circular orbit of radius r, with a constant angular velocity ω. Let T be the time period of revolution of the planet around the sun.

$$\omega = 2\pi/T$$

The centripetal force acting on the planet for its circular motion is:

$$F = mr\omega^2 = mr\,(2\pi/T)^2 = 4\pi^2 mr\,/\,T^2$$

According to Kepler's Third Law:

$$T^2 \, \alpha \, r^3 \text{ or } T^2 = K\,r^3$$

where K is a constant of proportionality.

Therefore,

$$F = 4\pi^2 mr/ Kr^3 = 4\pi^2 m / Kr^2 \quad \dots\dots\dots\dots\dots\dots\dots\dots\dots (1)$$

$$F \propto m/r^2 \text{ with } 4\pi^2/K \text{ as a constant}$$

This centripetal force is provided by the gravitational force between the sun and the planet. The gravitational force between the sun and the planet is mutual. If force F is directly proportional to the mass of the planet (m), it should also be directly proportional to the mass of the sun (M).

Hence, the factor

$$4\pi^2/K \propto M \text{ or } 4\pi^2/K = GM \quad \dots\dots\dots\dots\dots\dots\dots\dots (2)$$

where G is the proportionality constant which is now called the Gravitational constant.

Substituting eqn. (2) in the eqn. (1), we get:

$$F = G M m/r^2 \quad \dots\dots\dots\dots\dots\dots\dots\dots\dots\dots (3)$$

Eqn. (3) is Newton's Law of Gravitation.

The above law deduced by Newton is a mathematical expression of Kepler's third law relating the period and distance of a planet's orbit. From the first two laws, Newton had recognized the presence of force on the observed elliptical motion of planets. He also understood that force causes an object to accelerate.

Newton deduced the formula given in eqn. (3) for the force between the sun and a planet going around the sun. As a masterly stroke of genius, he also declared that the force that makes a planet go around the sun is of the same nature as the force that makes the apple fall to the ground. He thus asserted that eqn. (3) though deduced between the sun and a planet is also applicable for any two objects either on earth or anywhere in the cosmos.

The idea of gravity came to Newton when he observed the apple falling. However, to deduce the law of gravity he had to resort to Kepler's Laws of planetary motion. Had these laws not been available, Newton may not have had the law of gravity.

In 1686, Newton presented his book "The Principia" which contained his law of gravity to the Royal Society. Robert Hooke, an adversary of Newton, charged Newton with plagiarism claiming that Newton had obtained the Inverse Square Law from him. However, Newton was able to defend himself against this allegation of plagiarism by stating that he had already developed his theory of gravity before any significant interaction with Hooke. He was further able to prove that he was indeed the originator of the Inverse Square Law from a previous discussion he had with Sir Christopher Wren in 1679.

In 1798 (111 years after the publication of Newton's Principia and 71 years after Newton's death) the value of G was experimentally determined by Cavendish. Even though Newton had introduced G, he never knew the value of G during his lifetime.

Newton was troubled by many questions regarding his law of gravity, raised by contemporary scholars. For example, he could not provide a definite answer when questioned on the cause of gravity. It was a matter of common sense that one object can give force on another object only by physical contact. Then how can the Sun give force on a planet that is millions of miles away. After a lot of contemplation, Newton asserted that for gravity to act, physical contact between two objects is not necessary. As far as gravity is concerned, it acts at a distance without any medium between the objects - Gravity is "action at a distance".

Overall, Newton's law of gravitation has proven to be highly successful in describing and predicting the motion of celestial bodies. There were some irregularities in the orbit of Uranus that remained unexplained until 1846. It was John Couch Adams (1819-1892) in England and Urbain Jean Joseph Le Verrier (1811-1877) in France who independently used these irregularities to predict the existence and position of Neptune. The discovery of Neptune shortly thereafter was perhaps one of the most significant verifications of Newton's theory.

Even after about 350 years, Newton's law of gravity works and is extensively used in astronomy and astrophysics. All the required parameters of our satellites are calculated solely on Newton's law of gravity. This law remains a cornerstone of classical physics and a fundamental principle in our understanding of the universe.

However, for Newton, some aspects of his law nagged him throughout his life for lack of satisfactory explanations. Some of these aspects are:

a). Is the concept of action at a distance, correct?

Newton was not happy about his concept of action at a distance. He also could not point out a medium to transmit the gravitational force. Although Newton invented the concept of action at a distance to silence his critics, he eventually became his own critic. He expressed this sentiment in a letter to Richard Bentley: "That gravity should be innate, inherent, and essential to matter, so that one body may act upon another at a distance through a vacuum, without the mediation of anything else, by and through which their action and force may be conveyed from one to another, is to me so great an absurdity that I believe no man who has in philosophical matters a competent faculty of thinking can ever fall into it. Gravity must be caused by an agent acting constantly according to certain laws; but whether this agent be material or non-material, I have left open to the consideration of my readers."

b). What is the cause of gravity?

While mass is a measure of the strength of the gravitational force, Newton never said mass is the cause of gravity. In the year 1713 in the second edition of "Principia", he wrote: "I have not yet been able to discover the cause of these properties of gravity from phenomena and feign no hypothesis…. It is enough that gravity does really exist and acts according to the laws I have explained, and that it abundantly serves to account for all the motions of celestial bodies." Newton admitted that the cause of gravity was completely unknown to him. Despite several attempts, he could not locate the cause of gravity.

c). Is it possible to derive the law of gravity from basic axioms?

When Newton was pressed to derive his law by framing axioms, he refused to form any hypothesis declaring, 'Hypotheses non fingo', ("Principia" second edition, *1713)*. The question of deriving the law of gravity from basic axioms has remained a challenging problem since the days of Newton.

Why a Derivation?

All equations in Physics can be divided into two sets. Equations that are deduced from experimental observations and equations that are derived from basic concepts and axioms. Boyle deduced his law of gas (PV=constant) from experimental observations. Einstein derived his equation $E = mc^2$ from the axioms of special relativity.

For a simple pendulum, the expression $L / T^2 =$ constant (K) can be deduced from the experimental observations of L and the corresponding T values. The same expression can be derived from the kinematic considerations of the motion of the bob of the pendulum. The derivation gives an additional information that the constant K depends upon the value of the acceleration due to gravity (g) at the place of the pendulum and is given as $L/T^2 = K = g/4\ \pi^2$. Derivation leads to a more elaborate understanding of the system than the process of deduction.

Let's take another example of Boyle's gas law. Boyle deduced his gas law "PV = constant" from experimental observations. But when it was derived from the postulates of the kinetic theory of gases the law becomes

$$(P + a/V^2)\ (V - b) = RT$$

The above-derived expression provides more information than the deduced equation of Boyle.

Though Newton had explicitly stated that he did not want to form any hypothesis as to the cause of gravity, it is a historical fact that he did try to derive his law of gravity without using Kepler's Laws. In his attempt to derive, he introduced the existence of minute particles called fluxions. He gave fluxions a property of impinging and giving a thrust on any object in it. From this property, he tried to derive his formula but was not successful. However, his concept of fluxions was seriously taken up by Louis Le Sage, with some modifications. Around 1748, Le Sage developed a theory called Kinetic Theory of Gravity wherein the universe was assumed to contain a field of randomly moving tiny particles called ultra-mundane corpuscles with special assigned properties.

An isolated object A struck by these corpuscles equally from all sides results in an inward-directed pressure but no net directional force. With the presence of a second object B, a fraction of corpuscles that would have struck A from the direction of B, are intercepted. So, B works as a shield. Similarly, object A works as a shield for the corpuscles that would have normally struck B. Because of these shielding effects of different values, the corpuscles drive A and B towards each other. Thus, the apparent attraction between the bodies comes from the corpuscles pushing the objects towards each other. Later, the Kinetic Theory of Gravity came to be called Push Gravity or Le sage Gravity. Even though the term $(1 / r^2)$ was obtained in this theory, the complete law of gravity could not be obtained.

After Le Sage, a number of scientists like Nicholas Fatio de Duillier, George Louis Le Sage, de Broglie, Poincare, Majorana, George Darwin, Richard Feynman, tried to further develop the theory of Le Sage but were not successful. A variety of Le Sage models and related topics have been discussed in the book: Pushing Gravity: New Perspectives on Le Sage Theory of Gravitation by Edward in the year 2002.

Developing a theory of gravity from basic axioms, not from Kepler's laws, remains a challenge for scientists. This challenge is successfully addressed in detail under the Vethathirian Theory of gravity in the third chapter of this book.

Chapter II

Einstein and Gravity

In 1905, Einstein published his Special Theory of Relativity, unveiling remarkable outcomes derived from its fundamental axioms:

- Mass variation with velocity
- Time dilation
- Length contraction
- The universal speed limit: Nothing travels faster than light
- The iconic equation $E = mc^2$

The axioms that underpin these revelations are:

- The laws of physics remain consistent in all inertial frames.
- The velocity of light remains constant in all inertial frames.

Initially met with skepticism, Einstein's propositions gradually gained validation through successive experiments. However, a conflict emerged with Newton's notion of "action at a distance" as it contradicted Einstein's assertion that nothing surpasses the speed of light – a tenet challenged by the implied "infinite velocity" in action at a distance. Disturbed by this incongruity,

Einstein embarked on developing a new theory of gravity, aiming to eliminate the concept of action at a distance from the gravitational framework.

Einstein's Thought Experiments

In his pursuit of a new theory of gravity Einstein employed insightful thought experiments,

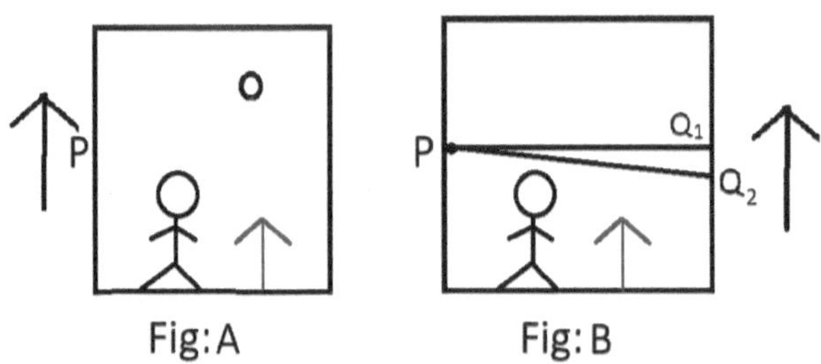

one of which involved an observer in an enclosed elevator (Fig. A) accelerating upward in a gravity free space. The observer inside the elevator is not aware of the fact that the elevator is being pulled up. He leaves a ball at a height. Now let us ask the question: Whether the ball will touch the bottom of the elevator or not? According to Einstein, the bottom of the elevator going up will meet the ball. His argument is the following: The moment the ball is placed, it is no longer accelerating, however, the elevator continues to accelerate upwards. So, the bottom of the elevator on its way up will meet the ball giving the impression that the ball has come down to the bottom of the elevator.

Thus, outside the elevator there is no gravity, but inside the elevator the observer will feel as if he is in a gravitational field. This shows that inside an accelerated frame, gravity can be created. From this experiment, Einstein posited that, despite the absence of gravity outside the elevator, the observer inside would experience a gravitational effect. This experiment, termed a thought experiment as it couldn't be physically executed, led to Einstein's groundbreaking Principle of Equivalence, where he asserted the equivalence of an accelerated frame and a gravitational field. He further described this assertion as the happiest thought of his life.

In a hypothetical scenario within the accelerating elevator, Einstein also considered the trajectory of a light pulse sent from one wall to the opposite (Fig. B). Consider a simple experiment conducted by the observer in the above elevator accelerating upwards. Let a light pulse be sent from a point P on one wall to the opposite wall across the elevator. Let Q_1 be the point on the opposite wall opposite to the point P, where the pulse was emitted. By the time the pulse reaches the opposite wall, Q_1 would have moved up a little bit and hence the pulse will hit a point Q_2 just below Q_1. The observer inside the elevator will say: The light pulse travels along a curved path. Einstein argued that, by the Principle of Equivalence, an observer in a gravitational field would undergo a similar experience, concluding that a light pulse would follow a curved path in a gravitational field. He attributed this curvature not to a force acting on the light pulse, but to mass curving the space the pulse traverses.

Einstein maintained that experiments carried out in an accelerating laboratory should yield equivalent results to those in a stationary laboratory within a gravitational field. This principle formed the basis of Einstein's General Theory of Relativity, which he presented in 1916 after delivering a series of four weekly lectures to the Prussian Academy of Sciences, describing it as "the most valuable work of my life".

Tensors

To know in general what is a tensor, let us consider the following situation.

Let $\vec{E}$ be an electrical vector of components (E^1, E^2, E^3) at a point in a metal. Let $\vec{C}$ be the current vector of components (C^1, C^2, C^3) produced at that point. Since $\vec{E}$ and $\vec{C}$ in the same direction, We write

$$\vec{C} = \sigma\vec{E} \quad \dots\dots\dots\dots\dots\dots\dots\dots\dots\dots\dots\dots\dots\dots\dots\dots \quad (1)$$

Here σ is a scalar and is called electrical conductivity of the metal.

Now let us repeat this process in a crystal.

Let $\vec{E}$ be an electric vector applied at a point in a crystal. Let $\vec{C}$ be the current produced. In a crystal $\vec{C}$ and $\vec{E}$ will not be in the same direction. In this case finding σ becomes a tedious process. What is done is the following:

E^1, E^2, E^3 are measured and C^1, C^2, C^3 are also measured by proper instruments, and terms like

$C^i / E^j = \sigma^{ij}$ $(i, j = 1,2,3)$ are determined. Now there are nine components like σ^{11}, σ^{12}, σ^{13}, σ^{21},........... σ^{33} for σ. A physical quantity having 9 components in a three dimensional space is called in general a tensor. Thus, electrical conductivity σ is a scalar in ordinary metal and a tensor with nine components in a crystal.

Transformation Property of Tensors

To know about transformation properties of tensor components, from one coordinate to another coordinate system let us consider a simple case of two-dimensional Cartesian coordinate X, Y with O as its origin.

Let x, y be the coordinates of a point P.

$$\overrightarrow{OP} = \vec{i}x + \vec{j}\,y \dots\dots\dots\dots\dots\dots\dots\dots \tag{2}$$

Let $(\overline{X},\ \overline{Y})$ be another coordinate system obtained by rotating X,Y) by an angle θ and $\vec{i}\,'$ and $\vec{j}\,'$ are unit vectors along $\overline{X}$ and $\overline{Y}$

$$\overrightarrow{OP} = \vec{i}\,'\,\overline{x} + \vec{j}\,'\overline{y} \dots\dots\dots\dots\dots\dots\dots\dots \tag{3}$$

$$\vec{i}\,'\,\overline{x} + \vec{j}\,'\overline{y} = \vec{i}x + \vec{j}\,y \dots\dots\dots\dots\dots\dots\dots\dots \tag{4}$$

where $\overline{x}$ and $\overline{y}$ are the new coordinates of the point P.

By taking a dot product of eqn. (4) with $\vec{i}\,'$ we get

$\overline{x} = \cos\theta\, x - \sin\theta\, y$

Similarly

$\overline{y} = \sin\theta\, x + \cos\theta\, y$

Since $\partial\overline{x}/\partial x = \cos\theta$ and $\partial\overline{x}/\partial y = -\sin\theta$

We can write,

$$\overline{x} = \left(\frac{\partial\overline{x}}{\partial x}\right)x + \left(\frac{\partial\overline{x}}{\partial y}\right)y \quad \text{and}$$

$$\overline{y} = \left(\frac{\partial\overline{y}}{\partial x}\right)x + \left(\frac{\partial\overline{y}}{\partial y}\right)y$$

Taking x as x_1 and y as x_2, we have

$$\overline{x}_1 = \left(\frac{\partial\overline{x}_1}{\partial x_1}\right)x_1 + \left(\frac{\partial\overline{x}_1}{\partial x_2}\right)x_2 \quad \text{and}$$

$$\overline{x}_2 = \left(\frac{\partial \overline{x}_2}{\partial x_1}\right) x_1 + \left(\frac{\partial \overline{x}_2}{\partial x_2}\right) x_2$$

Extending the above relation for n dimensional coordinate system, we write

$$\overline{x}_k = \left(\frac{\partial \overline{x}_k}{\partial x_1}\right) x_1 + \left(\frac{\partial \overline{x}_k}{\partial x_2}\right) x_2 + \left(\frac{\partial \overline{x}_k}{\partial x_3}\right) x_3 \ldots \left(\frac{\partial \overline{x}_k}{\partial x_n}\right) x_n$$

$$\overline{x}_k = \sum_{n=1}^{n} \left(\frac{\partial \overline{x}_k}{\partial x_n}\right) x_n = \left(\frac{\partial \overline{x}_k}{\partial x_n}\right) x_n \text{———————(5)}$$

where summation over any index is understood when it occurs twice in a term.

To sum up,

Let $(x^1, x^2 \ldots\ldots x^n)$ be the coordinates of a point P in one coordinate system of n dimension and $(\overline{x}^1, \overline{x}^2, \ldots\ldots \overline{x}^n)$ be the coordinates of the same point P in another coordinate system. Eqn. (5) shows the way by which the coordinates of a vector change from one system to another system.

In physics, there is a different kind of vector (for examples a $\vec{V}$ vector) whose components in two different coordinate systems are related as follows.

$$\overline{x}_k = \left(\frac{\partial x_n}{\partial \overline{x}_k}\right) x_n \ldots\ldots\ldots\ldots\ldots\ldots\ldots\ldots\ldots\ldots (6)$$

In this case, the vector is called a covariant tensor of rank 1. The index occurs as a subscript.

$$\text{When } \overline{x}^k = \left(\frac{\partial \overline{x}^k}{\partial x^n}\right) x^n \ldots\ldots\ldots\ldots\ldots\ldots\ldots\ldots\ldots\ldots (7)$$

the vector is called a contravariant tensor of rank 1. By convention, for such vector components the index occurs as a superscript.

Extending eqn. (6 and 7), we can write the transformation property for a tensor of rank 2 in a n dimensional coordinate system, as:

$$\overline{A}^{kj} = \left(\frac{\partial \overline{x}^k}{\partial x^n}\right) \left(\frac{\partial \overline{x}^j}{\partial x^i}\right) A^{ni} \ldots\ldots\ldots\ldots\ldots\ldots\ldots\ldots (8)$$

$$\overline{B}_{kj} = \left(\frac{\partial x_n}{\partial \overline{x}_k}\right) \left(\frac{\partial x_i}{\partial \overline{x}_j}\right) B_{ni} \ldots\ldots\ldots\ldots\ldots\ldots\ldots\ldots (9)$$

where A^{ni} is a contravariant tensor of rank 2 and B_{ni} is a covariant tensor of rank 2.

Metric Tensor

Let a particle move from a point (x^0, x^1, x^2, x^3) to $(x^0+dx^0, \ldots\ldots\ldots x^3+dx^3)$ with a velocity u. If the space - time interval between the above points is ds, from special Relativity we have for a flat space – time structure,

$$ds^2 = (dx^0)^2 - (dx^1)^2 - (dx^2)^2 - (dx^3)^2 \ldots\ldots\ldots\ldots\ldots (10)$$

The above expression for ds^2 in a flat space does not contain cross terms like $dx^1\, dx^2$, $dx^2\, dx^3$ etc. But in a curved space, the expression for ds^2 may contain cross terms with suitable coefficients. So, the most general expression for ds^2 will be,

$$ds^2 = g_{ij}\, dx^i\, dx^{j,} \ldots\ldots\ldots\ldots\ldots\ldots\ldots (11)$$

i and j run over 0,1,2,3

The metric tensor g_{ij} has sixteen components in a four dimensional frame. A physical entity having only one component (4^0) is called a scalar or tensor of rank 0. Examples are: Mass or temperature of a body. Velocity with four components in a four dimensional frame is a vector or a tensor of rank 1 (4^1). When we go from scalar to vector, the definition for the product of two vectors becomes complicated. When we go from flat space to curved space, differentiation becomes highly complicated. This is one reason why mathematical equations in curved space look complicated and tedious.

Tensor Differentiation

The process of differentiation becomes slightly complicated in a curved Space – Time geometry. If A^i is a component of a contravariant vector, its differentiation with respect to the co-ordinate x^k is given as

$$A^i_{;k} = \frac{\partial A^i}{\partial x^k} + \Gamma^i_{lk}\, A^l$$

where

$$\Gamma^i_{lk} = \frac{1}{2}\, g^{im} \left(\frac{\partial g_{mk}}{\partial x} + \frac{\partial g_{lm}}{\partial x^k} - \frac{\partial g_{kl}}{\partial x^m} \right)$$

If B_i is a component of a covariant vector then its differentiation with respect to $x^{k.}$

$$B_{i;k} = \frac{\partial B_i}{\partial x^k} - \Gamma_{ik}^{l} B_l$$

where, $g^{ij} = g_{ij} / det.g_{ij}$

Geodesics

Geodesics is the path along which a free particle moves in a Space – Time geometry. The Geodesics equation is of the form

$$\frac{d^2 x^\alpha}{ds^2} + \Gamma_{\mu v}^{\alpha} \frac{dx^\mu}{ds} \frac{dx^v}{ds} = 0$$

Schwarzschild Metric

In a four-dimensional curved space, the expression for ds^2 is given as,

$$ds^2 = g_{ij} dx^i dx^j \dots\dots\dots\dots\dots\dots(12)$$

The sixteen components of the covariant metric tensor g_{ij} of rank 2 will completely determine the magnitude of ds^2.

We discuss below the method by which Schwarzschild obtained all the sixteen components of g_{ij} for the space around a spherical body of mass m placed at the origin of a coordinate system.

Einstein's gravitational field equations are of the form

$$R_{ij} = k (g_{ij} T - T_{ij}) \qquad (13)$$

R_{ij} - Ricci tensor

T_{ij} = Energy–Momentum tensor of matter distribution

T is Energy– Momentum scalar

When the space outside the spherical body of mass 'm' is devoid of matter, T_{ij} = 0 and hence T=0, we have

$$R_{ij} = 0 \dots\dots\dots\dots\dots\dots\dots(14)$$

for the space around and outside the spherical body when there is no other object.

The space surrounding the body becomes non-Euclidean. In such a space the line element ds^2 is assumed to be

$$ds^2 = e^{\nu} c^2 dt^2 - e^{\mu} dr^2 - r^2 (d\theta^2 + \sin^2\theta \, d\phi^2) \dots\dots\dots\dots\dots\dots (15)$$

where ν and μ are functions of r and t. If $\nu = \mu = 0$, we get the flat space line element in spherical polar coordinates. The non- Euclidean effects are therefore contained in the functions μ and ν.

The Christoffel symbol and Ricci tensor R_{ij} which describes the nature of space around the mass are of the form,

$$\Gamma^i_{lk} = \tfrac{1}{2} g^{im} (\partial g_{mk} / \partial x^l + \partial g_{lm} / \partial x^k - \partial g_{kl} / \partial x^{m)}$$

$$R_{ij} = \partial \Gamma^a_{ij} / \partial x^a - \partial \Gamma^a_{ai} / \partial x^j + (\Gamma^a_{ab} \Gamma^b_{ij} - \Gamma^a_{ib} \Gamma^b_{aj})$$

Since $g_{00} = e^{\nu}$, $g_{11} = -e^{\mu}$, $g_{22} = -r^2$ and $g_{33} = -r^2 \sin^2\theta$, we have the following values for the Christoffel symbols.

$$\Gamma^1_{00} = \tfrac{1}{2} e^{\nu-\mu} \nu', \quad \Gamma^1_{11} = \tfrac{1}{2} \mu', \quad \Gamma^2_{33} = -\sin\theta \, \cos\theta,$$

$$\Gamma^1_{22} = -re^{-\mu}, \quad \Gamma^2_{12} = \tfrac{1}{r}, \quad \Gamma^3_{23} = \frac{\cos\theta}{\sin\theta},$$

$$\Gamma^0_{10} = \tfrac{1}{2} \nu', \quad \Gamma^1_{33} = -re^{-\mu} \sin^2\theta, \quad \Gamma^3_{13} = \tfrac{1}{r}$$

Using the above values for the Christoffel symbols, we obtain from Eqn. (14)

$$R_{00} = e^{\nu-\mu} \left\{ -\frac{\nu}{2} - \frac{\nu}{r} + \frac{\nu}{4} (\mu' - \nu') \right\} = 0$$

$$R_{11} = \frac{\nu''}{2} - \frac{\mu'}{r} + \frac{'}{4} (\nu' - \mu') = 0$$

$$R_{22} = \frac{R_{33}}{\sin^2\theta}$$

$$R_{33} = e^{-\mu} (1 - e^{\mu} + \frac{r}{2} (\nu' - \mu')) = 0$$

ν' or μ' indicate differentiation with respect to r

By adding $R_{(00)}$ and $R_{(11)}$ components, we get after suitable manipulations,

$$\nu (r) + \mu (r) = 0 \text{ or } \nu (r) = -\mu (r) \dots\dots\dots\dots\dots\dots\dots\dots (16)$$

From the expression for R$_{(22)}$, and by using the result v' = -μ', we get

$$1 - e^\mu - r\mu' = 0$$

$$(re^{-\mu})' = 1 \text{ or } e^{-\mu} = 1 + C/r$$

$$e^v = (1 + C/r) \dots\dots\dots\dots\dots\dots (17)$$

where C is an integration constant.

The term $(1 + C/r)$ shows that the space around the mass m has become non – Euclidean. In the derivation due to Schwarzchild, the constant C is taken as the -2m. In taking the constant C$_2$ as -2m, there is a kind of adhocism. In chapter - III we take a different route, more natural and logical for deriving Schwarzchild metric.

$$ds^2 = -\frac{dr^2}{\left(1 - \dfrac{2m}{r}\right)} - r^2\left(d\theta^2 + \sin^2\theta \; d\varnothing^2\right) + \left(1 - \frac{2m}{r}\right)c^2 dt^2 \dots\dots (18)$$

Planetary Orbits

In Comparison to sun of mass m the planets may be regarded as small free particles and their trajectories are given by geodesic equation

$$\frac{d^2x^\alpha}{ds^2} + \Gamma^\alpha_{\mu\nu}\frac{dx^\mu}{ds}\frac{dx^\nu}{ds} = 0 \dots\dots\dots\dots\dots (19)$$

$$\text{For } \alpha = 0, \quad \frac{d^2x^0}{ds^2} + \Gamma^0_{\mu\nu}\frac{dx^\mu}{ds}\frac{dx^\nu}{ds} = 0$$

$$\text{or } \frac{d^2t}{ds^2} + \Gamma^0_{10}\frac{dr}{ds}\frac{dt}{ds} + \Gamma^0_{01}\frac{dt}{ds}\frac{dr}{ds} = 0$$

$$\text{or } \frac{d^2t}{ds^2} + v'\frac{dr}{ds}\frac{dt}{ds} = 0 \dots\dots\dots\dots\dots\dots (20)$$

$$\text{For } \alpha = 1, \quad \frac{d^2x^1}{ds^2} + \Gamma^1_{\mu\nu}\frac{dx^\mu}{ds}\frac{dx^\nu}{ds} = 0$$

$$\text{or } \frac{d^2r}{ds^2} + \Gamma^1_{00}\left(\frac{dt}{ds}\right)^2 + \Gamma^1_{11}\left(\frac{dx^1}{ds}\right)^2 \Gamma^1_{22}\left(\frac{dx^2}{ds}\right)^2 + \Gamma^1_{33}\left(\frac{dx^2}{ds}\right)^2 = 0$$

$$\text{or } \frac{d^2r}{ds^2} + \frac{v'}{2}e^{v-\mu}\left(\frac{dt}{ds}\right)^2 + \frac{\mu'}{2}\left(\frac{dr}{ds}\right)^2 - re^{-\mu}\left(\frac{d\theta}{ds}\right)^2 - r\sin^2\theta \, e^{-\mu}\left(\frac{d\varnothing}{ds}\right)^2 = (21)$$

Further, for $\alpha = 2$, we get

$$\frac{dx^2}{ds^2} + \Gamma^2_{\mu\nu} \frac{dx^\mu}{ds} \frac{dx^\nu}{ds} = 0$$

or $\dfrac{d^2\theta}{ds^2} + \Gamma^2_{12} \dfrac{dx^1}{ds} \dfrac{dx^2}{ds} + \Gamma^2_{21} \dfrac{dx^2}{ds} \dfrac{dx^1}{ds} + \Gamma^2_{33} \dfrac{dx^2}{ds} \dfrac{dx^3}{ds} = 0$

or $\dfrac{d^2\theta}{ds^2} + \dfrac{2}{r} \dfrac{dr}{ds} \dfrac{d\theta}{ds} - r \sin\theta \cos\theta \left(\dfrac{d\varnothing}{ds}\right)^2 = 0$ . . . . (22)

Lastly for $\alpha = 3$, we have $\dfrac{d^2 x^3}{ds^2} + \Gamma^3_{\mu\nu} \dfrac{dx^\mu}{ds} \dfrac{dx^\nu}{ds} = 0$

or $\dfrac{d^2\phi}{ds^2} + \Gamma^3_{13} \dfrac{dx^1}{ds} \dfrac{dx^3}{ds} + \Gamma^3_{31} \dfrac{dx^3}{ds} \dfrac{dx^1}{ds} +$

$\qquad \Gamma^3_{23} \dfrac{dx^2}{ds} \dfrac{dx^3}{ds} \; \Gamma^3_{32} \dfrac{dx^3}{ds} \dfrac{dx^2}{ds} = 0$

or $\dfrac{d^2\phi}{ds^2} + \dfrac{2}{r} \dfrac{dr}{ds} \dfrac{d\phi}{ds} + 2 \cot\theta \dfrac{d\theta}{ds} \dfrac{d\phi}{ds} = 0$ . . . . (23)

We choose the coordinates such that the planet moves initially in the plane $\theta = \pi / 2$, so that $\sin\theta = 1$, $\cos\theta = 0$, $d\theta / ds = 0$. Substituting these values in the above equations.

$$\frac{d^2 t}{ds^2} + v' \frac{dr}{ds} \frac{dt}{ds} = 0 \qquad \text{. . . . . (24)}$$

$$\frac{d^2 r}{ds^2} + \frac{v'}{2} e^{v-\mu} \left(\frac{dt}{ds}\right)^2 + \frac{\mu'}{2} \left(\frac{dr}{ds}\right)^2 - r e^{-\mu} \left(\frac{d\varnothing}{ds}\right)^2 = 0 \text{ . . . . (25)}$$

$$\frac{d^2\phi}{ds^2} + \frac{2}{r} \frac{dr}{ds} \frac{d\phi}{ds} = 0 \qquad \text{. . . . . (26)}$$

Rewriting Eq. (26), we get

$$r^2 \frac{d^2\phi}{ds^2} + 2r \frac{dr}{ds} \frac{d\phi}{ds} = 0$$

or $\dfrac{d}{ds} \left(r^2 \dfrac{d\phi}{ds}\right) = 0$

Integrating this equation, we obtain

$$r^2 \frac{d\phi}{ds} = h \dots\dots\dots\dots\dots\dots\dots\dots\dots\dots\dots (27)$$

h is a constant of integration and is a measure of the angular momentum of the motion. Further, from Eqn. (20), we have

$$\frac{d^2t}{ds^2} + v' \frac{dr}{ds} \frac{dt}{ds} = 0$$

$$\text{Or } \frac{d}{ds} \left(e^v \frac{dt}{ds} \right) = 0$$

Integrating this equation, we get

$$e^v \frac{dt}{ds} = k \dots\dots\dots\dots\dots\dots\dots\dots\dots\dots\dots (28)$$

where k is the constant of integration. From Eqn.(15) we have,

$$1 = e^v \left(\frac{dt}{ds} \right)^2 - e^{-v} \left(\frac{dr}{ds} \right)^2 - r^2 \left(\frac{d\varnothing}{ds} \right)^2$$

$$\text{or } e^v \left(\frac{dt}{ds} \right)^2 - e^{-v} \left(\frac{dr}{ds} \right)^2 - r^2 \left(\frac{d\varnothing}{ds} \right)^2 - 1 = 0$$

Writing $\dfrac{dr}{ds} = \dfrac{dr}{d\varnothing} \dfrac{d\varnothing}{ds}$, and putting the values of $\dfrac{dt}{ds}$ and $\dfrac{d\varnothing}{ds}$ from Eqs. (28) and (27), we obtain

$$k^2 e^{-v} - \frac{h^2}{r^4} \left(\frac{dr}{d\varnothing} \right)^2 e^{-v} - \frac{h^2}{r^2} - 1 = 0$$

$$\text{or } h^2 \left(\frac{1}{r^2} \frac{dr}{d\varnothing} \right)^2 + \left(1 + \frac{h^2}{r^2} \right) \left(1 - \frac{2m}{r} \right) - k^2 = 0 \dots\dots\dots\dots\dots (29)$$

where we multiplied throughout by e^v.

Putting r = 1/u, one obtains

$$h^2 \left(\frac{du}{d\varnothing} \right)^2 + (1 + h^2 u^2)(1 - 2mu) - k^2 = 0$$

which on rearrangements is written as

$$\left(\frac{du}{d\varnothing}\right)^2 + u^2 = \frac{k^2-1}{h^2} + \frac{2m}{h^2}\,u + 2mu^3$$

Differentiating w.r.t. $\varnothing$, we obtain

$$2\,\frac{du}{d\varnothing}\,\frac{d^2u}{d\varnothing^2} + 2u\,\frac{du}{d\varnothing} = \frac{2m}{h^2}\,\frac{du}{d\varnothing} + 6\,mu^2\,\frac{du}{d\varnothing}$$

Dividing by $2\,\dfrac{du}{d\varnothing}$, we obtain

$$\frac{d^2u}{d\varnothing^2} + u = \frac{m}{h^2} + 3mu^2$$

$$\frac{d^2u}{d\varnothing^2} + u = \frac{m}{h^2} + 3mu^2 \dots\dots\dots\dots\dots\dots\dots\dots\dots(30)$$

The corresponding equation governing the orbit according to classical mechanics is

$$\frac{d^2u}{d\varnothing^2} + u = \frac{m}{h^2}$$

$$r^2\,\frac{d\varnothing}{dt} = h$$

Neglecting the term $3mu^2$ in the Eqn. (30), we obtain an approximate solution for u as

$$u = \frac{m}{h^2}\big(1 + e\cos(\varnothing)\big) \dots\dots\dots\dots\dots\dots\dots\dots(31)$$

Putting the above expression for u in the Eqn. (30), we get the solution for u as

$$u = \frac{1}{r} = \frac{m}{h^2}\left\{1 + e\cos\left(1 - \frac{3m^2}{h^2}\right)\phi\right\} \dots\dots\dots\dots\dots\dots\dots(32)$$

Let $\phi = 0$, correspond to a perihelion point P of the elliptical orbit of a planet around the sun.

$$u_p = \frac{1}{r_p} = \frac{m}{h^2}\{1+e\} \dots\dots\dots\dots\dots\dots\dots\dots\dots\dots\dots\dots(33)$$

Where r_p is the distance of P from the Sun.

u_p will have the same value m (1+e) / h² when $\left(1-\dfrac{3m^2}{h^2}\right)\phi = 2\pi$

or $\phi = 2\pi + 6\pi \ (\dfrac{m^2}{h^2})$

So, the next perihelion will not be at $\phi = 2\pi$ but at a different point where

$\phi = 2\pi + 6\pi\dfrac{m^2}{h^2}$. Thus, in one revolution the perihelion point advances

by an angle $\Delta\phi$ given as

$$\Delta\phi = 6\pi\frac{m^2}{h^2} \dots\dots\dots\dots\dots\dots\dots\dots\dots\dots\dots\dots(34)$$

Though for every planet its perihelion advances by $\Delta\phi$ in one revolution, it is appreciable and measurable for the planet mercury in the solar system. From eqn. (27) we write

$$r^2 \frac{d\phi}{ds} = h$$

Since ds = cdt, we have

$$\frac{r^2}{c}\omega = h \text{ where } \omega = \frac{d\phi}{dt}$$

$$\frac{r^4}{c^2}\omega^2 = h^2$$

Using the above relation we get

$$\Delta\phi = 6\pi\,\frac{m^2}{h^2}$$

$$= 6\pi\left(\frac{Gm}{c^2}\right)^2 \bigg/ \left(\frac{r^4\dot{u}^2}{c^2}\right)$$

G / c² is added here to make the right hand side to have proper dimension.

For the planet Mercury

$$\Delta\phi = 6\,\pi\left(\frac{Gm}{c^2}\right)^2 (ct)^2/r^4\ 4\pi^2$$

The number (n) of revolutions of the planet Mercury per century is:

$$n = 365 \times \frac{100}{T} = 36500/88 \text{ for Mercury}$$

$$= 414.7$$

The angle of advance in one century for the planet Mercury is:

$$\phi_M = (180 \times 3600/\pi)\quad 414.7 \times 6\,\pi\left(\frac{GM}{c^2}\right)^2 (ct)^2/r^4\ 4\pi^2$$

$$= 206369.4 \times 414.7 \times 6\,\pi\left(\frac{GM}{c^2}\right)^2 (ct)^2/r^4\ 4\pi^2 \ldots\ldots\ldots\ldots(35)$$

for the elliptical orbit of any planet, we have

$$(ct)^2 = \frac{4\pi^2}{\left(\frac{Gm}{c^2}\right)}\,r^3 \text{ or } \left(\frac{Gm}{c^2}\right) = \frac{4\pi^2}{c^2 T^2}\,r^3 \ldots\ldots\ldots\ldots\ldots\ldots\ldots(36)$$

Combining Eqn.(35) and (36), We have

$$6\pi\,\frac{\left\{\dfrac{16\pi^4}{c^4 t^4}\cdot r^6\right\}(tc)^2}{4\pi^2 r^4} = \frac{24\ \pi^3\ r^2}{c^2 t^2} = 43 \text{ seconds.}$$

The advance for Mercury perihelion, is known from the time of the astronomer Leverrier (1859). It may be remarked that Newton's theory of gravitation also predicts the advance of the perihelion of any planet due to perturbations caused by other planets and several other causes. However, after accounting for all these effects, there still remains an effect of about 43 seconds per century which cannot be accounted for by Newton's theory. This is now accounted for by Einstein's theory.

Now we quote Einstein: "After taking account of all the disturbing influences exerted on Mercury by the remaining planets, it was found (Leverrier: 1859; and Newcomb; 1895) that an unexplained perihelian movement of the orbit of Mercury remained over, the amount of which does not differ sensibly from the above mentioned +43 seconds of arc per century".

Deflection of Light Ray in Gravity

Since light has gravitating mass, so a ray of light on passing near the sun will deviate from its original path. Actually, J. Soldner in 1801, calculated that a star viewed near the sun, would be displaced by 0.88 second on the basis of Newtonian theory. Let us treat the deflection of a light ray on the basis of Einstein's theory.

The exact relativistic equation governing the shape of all photon orbits can be obtained from Eqn.(30) as

$$d^2u \,/\, d\phi^2 + u = 3mu^2 \dots\dots\dots\dots\dots\dots\dots\dots(37)$$

In the Eqn. (27), h becomes infinite since ds becomes zero for light and hence m/h^2 becomes zero.

The solution for the above equation is of the form,

$$u = \frac{sin\phi}{R} + \frac{3m}{2R^2}(1+\frac{1}{3}cos2\phi) \dots\dots\dots\dots\dots\dots(38)$$

For large values of r, ϕ becomes extremely small so that $\sin\phi$ becomes ϕ and $\cos 2\phi$ becomes 1. In this case, taking the value of ϕ as α, for large value of r we have.

$$0 = \frac{\phi}{R} + \frac{3m}{2R^2}\left(1+\frac{1}{3}\right)$$

$$0 = \frac{\phi}{R} + \frac{2m}{R^2}$$

$$0 = R\,\phi + 2m$$

$$\phi = -\frac{2m}{R}$$

Hence the magnitude of the total deflection (ϕ_M) of the ray will be 4m/r, or 4Gm/ C^2R.

For the case of sun, m $= 1.99 \times 10^{33}$ gm, R $= 6.97 \times 10^{10}$ cm, so

$\phi_M = 1.75$ arcseconds

The above results shows that a ray of light passing at grazing incidence, near a gravitating mass like sun, will be deflected by an angle 1.75sec. Observations about the passage of the light rays from the stars behind the sun grazing the surface of the sun and reaching the earth were taken during the total eclipse of the sun. By studying the apparent positions of the stars whose light has passed near the sun at grazing incidence, it has been concluded that the theoretical prediction of Einstein agrees with the value of the experimental observation, thus validating the prediction of the General Theory of Relativity regarding the bending of light in a gravitational field.

The bending angle is indeed very small; its measurement was done by Eddington and his colleagues in 1919 at the time of a solar eclipse.

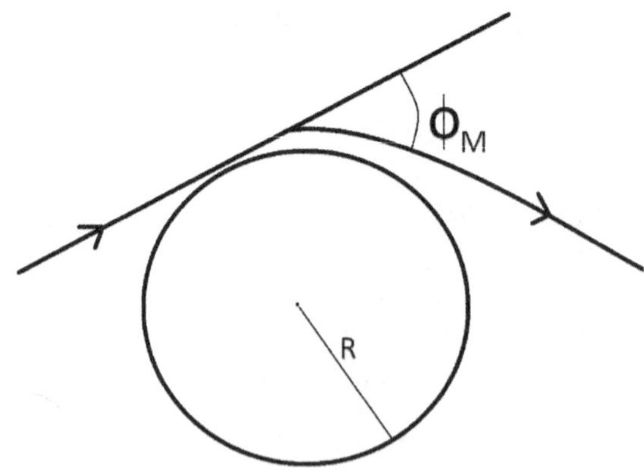

Let us quote Einstein: "We are indebted to the (British) Royal Society and to the Royal Astronomical Society for the investigation of this important deduction. Undaunted by the First World War and by difficulties of both a material and a psychological nature aroused by the war, these societies equipped two expeditions – to Sobral (Brazil), and to the island of Principe (West Africa) – and sent several of Britain's most celebrated astronomers (Eddington, Cottingham, Crommelin, Davidson), in order to obtain photographs of the solar eclipse of 29th May, 1919. The relative discrepancies to be expected between the steller photographs amounted to a few hundredths of a millimetre only. Thus great accuracy was necessary in making the adjustments required for the taking of the photographs and in their subsequent measurement. The Results of the measurements confirmed the theory in a thoroughly satisfactory manner".

Gravitational Red-shift of Spectral Lines

$$f_0 / f = (1 + 2Us / c^2) / (1 + 2U_e / c^{2)} \dots\dots\dots\dots\dots\dots\dots\dots (39)$$

where f is the frequency of a spectral line emitted by an element on the sun as measured in a laboratory on earth and f_0 is the frequency of the same spectral line emitted by the same element on the earth as measured in the same laboratory.

where U_s and U_e are the gravitational potential of the sun on its surface and on the surface of the earth.

Since U_e is greater than U_s, f_0 will be less than f. Thus when the spectral line travels from the sun to the earth, its frequency decreases. This is interpreted by Einstein in the following way.

The frequency of an atom of an element situated on the surface of the sun, as measured by an observer on the earth will be somewhat less than the frequency of the same atom of the same element which is situated on the earth.

$$\text{Since } U_s = -1.914 \times 10^{11}, = -1914 \times 10^8$$
$$U_e = -9.512 \times 10^8 = -9.512 \times 10^8$$
$$f_0 = 0.9999979f$$

Thus f_0 differs from f by an extremely small quantity. Einstein's remark about this situation is the following: It is an open question whether or not this effect exists, and at the present time (1920) astronomers are working with great zeal towards the solution. Owing to the smallness of the effect in the case of the sun, it is difficult to form an opinion as to its existence.

At all events, a definite decision will be reached during the next few years. If the displacement of spectral lines towards the red by the gravitational potential does not exist, then the general Theory of Relativity will be untenable.

The above predicted effect is about thirty times larger for a spectral line emanating from the companion of Sirius and the effect has been confirmed by observing the spectral lines coming from it.

Conclusion

Albert Einstein's groundbreaking contributions to science, particularly in physics, are epitomized by the concepts of space-time unification and space-time curvature. Unlike Newton, who considered space passively, Einstein endowed it with a dynamic role, leading to profound implications.

In Einstein's gravitational theory, Galileo's principle is a fundamental element, integral to its coherence. Einstein envisioned a deep connection between inertia and gravity, positing that inertial and gravitational forces are essentially identical. He emphasized that while this equivalence was acknowledged in mechanics, it lacked a satisfactory interpretation.

Einstein's theory of gravity revolutionized our understanding, notably evidenced by the deviation in the perihelion point of planetary orbits from Newton's predictions. Additionally, his theory accurately predicted phenomenon such as the bending of light in a gravitational field and gravitational red-shift of the spectral lines from a star when viewed from earth. For these effects, Einstein's theory predicted better values than the values given in Newton's and these were confirmed by experiments.

Einstein's framework, which treats every natural event as a point in a four-dimensional coordinate system, has become a cornerstone in physics. It successfully explains phenomena ranging from black holes and orbiting neutron stars to gravitational lensing, as well as the relativistic time dilation effect observed on Earth's surface compared to space.

We admire the preeminent genius of Einstein in linking gravity with the metric tensor of a suitably chosen space-time coordinate system. However, we should mention that his theory of gravity had sidestepped the challenges mentioned in Chapter III, imposing a cost when addressing the universe on a cosmological scale. Initial cosmological concepts presented by Einstein in 1920 were later debunked by Hubble's experimental observations in 1928. Einstein's unaddressed challenges persisted until his final days in 1954, underscoring the need to scrutinize and resolve these issues for a comprehensive understanding of cosmology. In the subsequent chapter, we delve into these unresolved challenges, introducing a new theory of gravity to address them.

Chapter III

Vethathirian Concept of Gravity

In earlier chapters, we saw that Newton's law of universal gravitation is useful in everyday situations. However, Einstein's general theory of relativity is more accurate for extreme conditions of high gravitational field and when dealing with cosmological phenomena. However, we also know that both of these gravity models have some serious limitations. The Vethathirian theory of gravitation is a new way to explain gravity. It considers Space as the only independent entity with time, matter and energy all emanating from it. The Vethathirian concept about Space provides a crucial foundation for understanding gravity and the anti-gravity dark energy.

Einstein once remarked "I have little patience with scientists who take a board of wood, look for its thinnest part, and drill a great number of holes where drilling is easy." When Einstein made his remark, he was expressing his impatience with scientists who approach their work in a simplistic or reductionist manner.

The work presented in this chapter required us to approach in an exact opposite way. We have taken a more holistic and comprehensive approach when addressing the fundamental unresolved issues in physics that have been with us for centuries. Through the application of the Vethathirian gravity model, we've effectively addressed these intricate challenges and queries listed below. The content presented in this chapter holds significant importance and lays the groundwork for the development of increasingly coherent and comprehensive cosmological models in the future.

- What is the cause of gravity?
- What is the mechanism by which one body exerts a force on the other?
- Can we theoretically prove Galileo's demonstration that two objects with different masses fall at the same speed?
- How and where does an object acquire its massiveness?
- Is there a derivation for Newton's law of gravity from first principles?
- What is the mechanism by which a mass curves space?
- What is the source of Matter and the antigravity Dark energy?

Newton had himself acknowledged that his idea of "action at a distance" is counterintuitive and not logical enough. This is because there is no clear explanation for how an object can exert a force on another object that is millions of miles away. Similarly, in Einstein's concept of gravity, there is no proper mechanism to explain how a mass is able to curve a boundaryless Space. The concept of Space-Time curvature in Einstein theory of gravity is more of mathematical in character without due physical insight. The concept of Space-Time curvature in Einstein's theory of gravity is fundamentally rooted in intricate mathematical formulations of Riemann geometry. In this chapter we also review the concepts of "action at a distance" and the "curvature of space" in more detail.

Framing Axioms for Space

In physics, we know that Newton talked of four separate entities: Space, Time, Matter, and Energy. Einstein reduced them to just two entities: Space-Time and Matter-Energy. According to the Vethathirian model of gravity, Space stands as a singular, primordial entity that gives rise to time, matter, and energy.

Einstein reduced the four independent entities of Newton to two and we reduce the two entities of Einstein to one and that one is Space. Vethathirian concept of gravity begins with the following two axioms for Space (Primordial Space):

- Space is all-pervading and is of potential Energy and Consciousness. It has the property of constant self-compression, and of continually exerting compressive pressure on every system in it.

- Self-compression results in the formation of infinitesimal spinning quanta of Space, called "formative dust." Due to the surrounding compression, dust are forced into the formation of discrete groups called fundamental particles. Every group of dust formed by the surrounding compressive pressure of Space has a spin and hence becomes a source of a diverging radial field with a repulsive force at every space point.

The first axiom considers Space as full of potential Energy. This Energy is a real thing, not like the "negative energy sea" concept of Dirac or "the zero-point energy" of a quantized field which are purely mathematical. Space exists even before the formation of any field or any material particle.

By the first axiom, we postulate that Space exerts pressure on any two particles within it, causing them to draw closer. Hence for two particles coming closer, Space is the underlying cause. Take the example of the falling apple; it is neither the earth nor the apple that initiates this motion. Rather, it is the Space surrounding the earth and the apple pushing them together. The compressive property of Space brings them closer and closer. Thus, Space is the cause for gravity, which is the first-born force in the universe.

We acknowledge that our viewpoint on the source of gravity diverges from the prevailing notion that the earth attracts the apple. It's worth noting that this traditional idea of a particle attracting another particle has not helped us to derive the law of gravity from basic axioms. However, with our position that particles are pushed toward each other by Space, we demonstrate that we can actually derive Newton's law of gravity, without using Kepler's laws of planetary motion.

Our second axiom posits that primordial Space serves as the source of fundamental particles. It also proposes a mechanism for the generation of diverging waves in Space. These waves, characterized by their repulsive (anti-gravity) nature, bear resemblance to the cosmological constant initially introduced by Einstein. Consequently, this axiom introduces a repulsive force that could be referred to as a fifth force, operating on a cosmic scale.

We are of the opinion that this fifth force may be the elusive dark energy that contemporary cosmologists grapple with – a phenomenon considered one of the most perplexing enigmas in modern astrophysics and cosmology, eluding easy comprehension.

Newton considered matter as existing forever and gave a similar picture of permanence to energy as well. Einstein extended this notion by considering the combined matter and energy content of the universe as a constant over time. However, neither of them delved into the source of matter and energy. In this context our second axiom assumes profound significance. By positing matter as a product of Space, this axiom goes far beyond the perspectives of both Newton and Einstein. Our concept that Space is the source of matter, energy and time marks a significant leap forward in our understanding of the universe and its contents.

Through our axioms of Space, we have expanded the realm of physics from a particle-based model to a Space based one, recognizing primordial Space as the wellspring of particles, gravity, and dark energy. With this expansion, physics undergoes a fundamental transformation, evolving into a 'new physics' that redefines the very essence of this discipline. Below, we demonstrate how this redefined physics provides insight into the nature of mass. The formula for the mass of an object derived from its causative factors along with a derivation of the law of gravity, both discussed below, emerge as immediate outcomes of our Redefined Physics.

A Formula for Mass

Since the law of gravity contains in it the mass of a body, we need to understand "what is mass" so as to derive the law of gravity.

As mentioned in our axioms, Space has the property of exerting a compressive pressure on every object in it. Let C be the compressive pressure on a spherical body due to Space. Then the total compressive thrust on the body will be AC, where A is the surface area of the body.

Because the constituent particles in the spherical body spin, by our second axiom, there will be a repulsive pressure which we call R emanating from each unit area of the body. The total repulsive thrust emanating from the body will be AR. The net compressive thrust or the net gripping thrust on the body due to Space will then be A(C − R).

As a result of this net compressive thrust, the body is in the grip of Space. Because of this grip, an external force applied on the body finds it difficult to move it. The cause for this difficulty is not the body but the Space. The stronger the gripping thrust on the body, the more difficult it will be to move the body. This difficulty is traditionally interpreted as the massiveness of the body or the inertia of the body. When we say that a body is massive, we forget that Space is the source of that massiveness. This has been a costly slip in science, which continues even today.

It is apparent that the massiveness or the inertial mass (M) of a body will be proportional to the net gripping thrust on the body due to Space. Hence, we can write,

$$M \propto A\,(C - R) \text{ or}$$
$$M = \beta\,A\,(C - R) \dots\dots\dots\dots\dots\dots\dots\dots\dots\dots\dots \quad (1)$$

where β is a universal constant, whose dimension is $L^{-1}T^2$

If we apply a force F on M the acceleration (a) produced will be

$$a = F/M = F/\,\beta\,A\,(C - R) \dots\dots\dots\dots\dots\dots\dots\dots\dots \quad (1a)$$

The above mass formula is named Alagar – Uma mass formula in Vethathirian gravity model and clearly reveals that the mass of an object is caused by the actions of C and R. The key insight or the message here is that the mass of a body is not inherent to the body itself but is acquired by the body due to $(C - R)$.

The mass of an object can now be defined as a measure of the net compressive thrust exerted on it by Space.

In Vethathirian gravity, there is only one mass for an object given by the eqn. (1) and that is called the inertia of the object. Since gravity is taken as a property of Space, there is no need to define a gravitational mass for an object with a property to attract other masses. In both Newton and Einstein's theories, an object is assumed to have two masses: m_i and m_g, m_i being the inertial mass and m_g being the gravitational mass. Galileo's Pisa tower experiment demands that m_i be equal to m_g. But till today, there has been no theoretical reason given as to why the two masses should be equal.

Vethathirian Theory of Gravity speaks only one mass for an object, that too, not inherent but acquired by the object due to the action of Space on it.

This one mass may be considered as what Einstein wished for when he wrote in his book: "A satisfactory interpretation (of Galileo's, demonstration) can be obtained only if we recognized the following fact: The same quality of a body manifest itself according to circumstances as inertia or as weight".

The equation $M = \beta A(C - R)$ is the first ever equation in physics linking matter and boundaryless Space. In the above equation, C is a factor representing the action of Space on matter and R is the factor representing the action of the dark energy repulsive pressure, opposing C.

Our mass formula, derived from our axioms, can be used to derive the law of gravity without using Kepler's Laws. This derivation shown below demonstrates both the power of the mass formula and the power of our axioms.

Law of Gravity Finally Derived

A derivation of the law of gravity from basic axioms has been pending in physics since the days of Newton. To quote Professor Brian Cox: "Newton's gravitational constant is one of the fundamental physical constants. It describes a property of our universe that can be measured but not derived from some deeper principle, as far as we know."

We present below a derivation of the law of gravity developed by Alagar Ramanujam.

Let M_1 be the mass of a spherical body A of surface area A_1 placed at the origin of a coordinate system. (Fig.1)

From our first axiom, Space will converge on M_1 constituting a flow of thrust towards M_1. Whatever thrust flows through a spherical surface of radius r will also flow through any other spherical surface of radius less than r or greater than r.

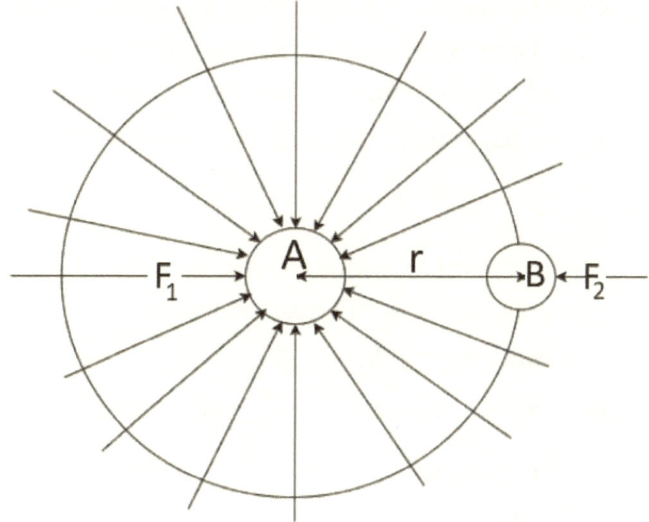

(Fig.1)

By our mass formula, we have

$$M_1 = \beta \, A_1 \, (C_1 - R_1) \quad \dots\dots\dots\dots\dots\dots\dots\dots\dots\dots\dots\dots\dots (2)$$

where C_1 is the compressive pressure on the body and R_1 is the repulsive pressure emanating from unit area of the surface of the body A.

The total compressive thrust on the body A due to Space is $A_1 C_1$. The total compressive thrust passing through a sphere of radius r will also be $A_1 C_1$. The compressive pressure on the unit area of the sphere of radius r will be $A_1 C_1 / 4\pi r^2$. (Fig.1)

Since R_1 is the repulsive flux through unit area of the body A, the total repulsive thrust from the body A is $A_1 R_1$. This total thrust will cross the sphere of radius r and hence the repulsive pressure on unit area of the sphere of radius r will be $A_1 R_1 / 4\pi r^2$.

The net compressive pressure on unit area at the distance 'r' from the body A will be $A_1 (C_1 - R_1) / 4\pi r^2$

Let us now place a spherical body (B) of mass M_2, at a distance r from the center of body A. The compressive force (F_2) on B can be written as

$$F_2 = K_2 \, A_1 \, (C_1 - R_1) / 4\,\pi r^2 \quad \dots\dots\dots\dots\dots\dots\dots\dots\dots\dots\dots (3)$$

where K_2 will be a function of M_2 with a dimension of area. K_2 is the effective area of the body B exposed to the flow of thrust towards A. Here Space gives the force F_2 on the body B trying to move it towards A.

By symmetry, we can write the force (F_1) acting on the body A,

$$F_1 = K_1 A_2 (C_2 - R_2) / 4\,\pi r^2 \dots\dots\dots\dots\dots\dots\dots\dots \quad (4)$$

where K_1 will be a function of M_1 with a dimension of area. K_1 is the effective area of body A exposed to the flow of thrust towards B.

By our mass formula, $M_2 = \beta A_2 (C_2 - R_2)$

According to Newton's third law, $F_1 = F_2$, and therefore,

$K_1 / K_2 = A_1 (C_1 - R_1) / A_2 (C_2 - R_2)$
Let $K_1 = \mu (r, t) A_1 (C_1 - R_1)$ and
$K_2 = \mu (r, t) A_2 (C_2 - R_2)$
where $\mu (r, t)$ is a function of r and t and its dimension is $LM^{-1} T^2$

$$F_1 = F_2 = \mu (r, t) A_2 (C_2 - R_2) A_1 (C_1 - R_1) / 4\pi r^2 \dots\dots\dots\dots\dots\dots (5)$$

The most significant aspect of eqn. (5) is that the gravity force between two bodies A & B is expressed purely in terms of space parameters. We can also express eqn. (5) in terms of the masses M_1 and M_2 of the spheres A and B by using our mass formula given in eqn (1). Thus, we have,

$$F_1 = F_2 = \mu (r,t) M_1 M_2 / 4\pi r^2 \beta^2$$
$$= G(r,t) M_1 M_2 / r^2 \dots\dots\dots\dots\dots\dots\dots\dots\dots\dots\dots (6)$$
where $G(r,t) = \mu(r,t) / 4\pi \beta^2$ and is of dimension $L^3 M^{-1} T^2$

Thus, we derive Newton's law of gravity without using Kepler's Laws. There is now a sense of satisfaction that the problem of deriving Newton's law of gravity from first principles has been finally solved here.

It must be noted here that the mass occurring in eqn. (1a) determining the acceleration produced by an applied force and the mass occurring in eqn. (6) determining the magnitude of the gravitational force between two bodies is of the same quality. There is no need in our theory to introduce two separate masses m_i and m_g called inertial and gravitational masses for a given body.

The derivation of a well-known formula by a fresh set of axioms has great significance. It demonstrates not only the validity of the formula but also offers a deeper insight into the terms occurring in the formula. For example, in Newton's gravity formula G is the proportionality constant. In Einstein's General Theory, G enters as an integration constant. But in our derivation, G is a product of two factors and can be, in general, a constant or a function of r and t. When Newton deduced the gravity formula, he treated the mass of an object as an inherent property of the object. But, in our formalism the massiveness of an object is a consequence of the net compressive grip on it due to Space and hence not an inherent property of the object. The net compressive grip mentioned above may vary with time.

Since Space is invisible and was considered as vacuum in the days of Newton, people were tempted to conclude that M_1 gives the force on M_2 and M_2 gives force on M_1 Making the two masses to move towards each other and hence they declared that a particle attracts another particle. As shown here, it is the Space that gives forces on A and B trying to bring them together and thus we establish that gravity is a property of Space.

Eqn. (6) leads us to an important conclusion discussed below. The acceleration (a_2) of the mass M_2 towards M_1 can be obtained as

$$a_2 = F_2 / M_2 = G\, M_1 / r^2$$

The value of a_2 is determined only by M_1 and not by M_2. Whatever the value of M_2, big or small its acceleration a_2 towards M_1 will be the same. This fact which emerges from our derivation was experimentally demonstrated by Galileo several centuries ago. Einstein used this result of Galileo's demonstration to build his theory of gravity. We are happy that a theoretical proof for the result demonstrated by Galileo and used by Einstein, emerges from our derivation of the law of gravity.

Action at a Distance

A significant difference between the Newtonian concept of gravity and the Vethathirian concept of gravity is discussed below. Even though the final equation for the gravitational force between two masses is the same, both in Newtonian and Vethathirian concept of gravity our derivation process has provided certain additional insights. We have,

$$F = G\, M_1 M_2 / r^2$$

According to Newton, in the absence of the second mass M_2, the gravitational force F is zero everywhere. Left alone, the mass M_1 does not produce any change in its surroundings. However, in Vethathirian concept, it is not so. The moment M_1 is placed, Space starts to converge on it. In addition, the dark energy repulsive flux leaves the surface of M_1 in all directions. Thus, there is a gravitational pressure field $(C_r - R_r)$ at every point around M_1 even when M_2 is not there. Here 'r' is the distance of the point from the center of the mass, M_1. Thus, the presence of M_1 has changed the very nature of the Space around it. The Space around M_1 has become a gravitational field even in the absence of M_2. The second mass M_2 is required only to demonstrate the existence of that field.

Due to the presence of M_1 the space around it gets divided into a series of concentric spheres (or nested spheres). On a unit surface area of a particular sphere of radius r there will be an inward pressure field $(C_r - R_r)$ directed towards M_1. As the space is converging on M_1, $(C_r - R_r)$ goes on increasing as we move towards M_1.

If we place a small test body at a distance r from M_1, the test body will experience a force whose value will be proportional to the inward pressure $(C_r - R_r)$. This force acts on the test body instantly upon placement. This implies that nothing travells from the first spherical body to the test body, with an infinite velocity that concerned Einstein very much. Since the force acts immediately on placement, it's not a case of action at a distance, but rather an action occurring right then and there. As Newton had completely neglected the role of Space, he had no alternative but to assume that a kind of influence jumped from one body to another, with an infinite velocity resulting in a force that brings them closer together.

Space – Time Curvature and the Schwarzschild Metric

As previously discussed, with the presence of a spherical body of mass M, at an origin O, the Space around it becomes a field with concentric spheres (or nested spheres), with the origin as the center. Each sphere is characterized by a unique value for (C_r-R_r) where r represents the radius of the sphere. As Space converges more and more towards the body, the total inward thrust $4\pi r^2$ $(C_r - R_r)$ will remain the same on all the concentric spheres. As we approach the body, the area of the spheres decreases and the net inward pressure $(C_r - R_r)$

acting on the sphere increases. It finally reaches a maximum value on the surface of the spherical body of radius R.

We could consider the division of Space into a series of concentric spheres (or nested spheres) by the presence of a mass as an underlying mechanism for the concept of Space curvature introduced by Einstein. In a distant way, formation of the concentric spheres can be interpreted as the spherical body distorting Space or bending Space.

At each and every unit area in Space, the nature of curvature is decided by $(C - R)$ at that point. The higher the value of $(C - R)$, the higher will be the curvature at that point. The total thrust on the sphere of radius r will be $4\pi r^2$ $(C_r - R_r)$ and this remains the same for all spheres whatever the radius. The mechanism discussed here offers a deep insight into the concept of curvature introduced by Einstein. Using this mechanism, we present below an alternative derivation of the Schwarzschild metric.

Let us recall the derivation of Schwarzschild metric relation given in the second chapter. A short version of the derivation is given below.

Schwarzschild Metric

In a four-dimensional curved Space-Time structure as discussed in the earlier chapter, the expression for ds^2 is given as,

$$ds^2 = g_{ij}\, dx^i\, dx^j \dotfill (7)$$

The sixteen components of the covariant metric tensor g_{ij} of rank 2 will completely determine the magnitude of ds^2.

We discuss below the method by which Schwarzschild obtained all the sixteen components of g_{ij} for the space around a spherical body of mass 'm' and of radius 'R' placed at the origin of a coordinate system.

Einstein's field equations for gravitation are of the form

$$R_{ij} = k\, (g_{ij}\, T - T_{ij}) \dotfill (8)$$

R_{ij} - Ricci tensor

T_{ij} = Energy–Momentum tensor of matter distribution

T is Energy – Momentum scalar

When the space outside the spherical body of mass 'm' is devoid of matter, T_{ij} = 0 and hence T=0, we have

$$R_{ij} = 0 \dots\dots\dots\dots\dots\dots\dots\dots\dots\dots\dots\dots\dots\dots\dots\dots (9)$$

for the Space around and outside the spherical body.

For the Space surrounding the spherical body, the line element ds^2 is assumed to be

$$ds^2 = e^{\nu} c^2\, dt^2 - e^{\mu}\, dr^2 - r^2\, (d\theta^2 + \sin^2\theta\, d\phi^2) \dots\dots\dots\dots (10)$$

where ν and μ are functions of r and t. If $\nu = \mu = 0$, we get the flat space line element in spherical polar coordinates. The non- Euclidean effects are therefore contained in the functions μ and ν.

The Christoffel symbol and Ricci tensor R_{ij} which describes the nature of space around a mass are of the form,

$$\Gamma^i_{lk} = \frac{1}{2}\, g^{im} (\partial g_{mk} / \partial x^l + \partial g_{lm} / \partial x^k - \partial g_{kl} / \partial x^{m)}$$

$$R_{ij} = \partial \Gamma^a_{ij} / \partial x^a - \partial \Gamma^a_{ai} / \partial x^j + (\Gamma^a_{ab}\Gamma^b_{ij} - \Gamma^a_{ib}\Gamma^b_{aj})$$

Since $g_{00} = e^{\nu}$, $g_{11} = -e^{\mu}$, $g_{22} = -r^2$ and $g_{33} = -r^2\sin^2\theta$, we have the following values for the Christoffel symbols.

$$\Gamma^1_{00} = \frac{1}{2}\, e^{\nu-\mu}\nu', \quad \Gamma^1_{11} = \frac{1}{2}\, \mu', \quad \Gamma^2_{33} = -\sin\theta\,\cos\theta,$$

$$\Gamma^1_{22} = -r e^{-\mu}, \quad \Gamma^2_{12} = \frac{1}{r}, \quad \Gamma^3_{23} = \frac{\cos\theta}{\sin\theta},$$

$$\Gamma^0_{10} = \frac{1}{2}\, \nu', \quad \Gamma^1_{33} = -r e^{-\mu}\sin^2\theta, \quad \Gamma^3_{13} = \frac{1}{r}$$

Using the above values for the Christoffel symbols, we obtain from Eqn. (9)

$$R_{00} = e^{\nu-\mu} \left\{ -\frac{\nu''}{2} - \frac{\nu'}{r} + \frac{\nu'}{4}(\mu' - \nu') \right\} = 0$$

$$R_{11} = \frac{\nu''}{2} - \frac{\mu'}{r} + \frac{\nu'}{4}(\nu' - \mu') = 0$$

$$R_{22} = e^{-\mu}\{ 1 - e^{\mu} + \frac{r}{2}(v' - \mu')\} = 0$$

$$R_{33} = \sin^2\theta \; R_{22} = 0$$

v' or μ' indicate differentiation with respect to r

By adding R_{00} and R_{11} components, we get after suitable manipulations,

$$v(r) + \mu(r) = 0 \text{ or } v(r) = -\mu(r). \dots\dots\dots\dots\dots\dots (11)$$

From the expression for R_{22}, and by using the result $v' = -\mu'$, we get

$$1 - e^{\mu} - r\mu' = 0$$

$$(re^{-\mu})' = 1 \text{ or } e^{-\mu} = 1 + C/r \dots\dots\dots\dots\dots\dots (12)$$

$$e^{v} = (1 + C/r) \dots\dots\dots\dots\dots\dots\dots (13)$$

where C is an integration constant.

The term $(1 + C/r)$ shows that the space around the mass m has become non – Euclidean. In the derivation due to Schwarzschild, the constant C is taken as $(-2m)$. In taking the constant C as $-2m$, there is a kind of adhocism. We take here a different route for deriving Schwarzschild metric which is more straight forward and logical.

As discussed already, with the presence of mass m, the space around it gets divided into concentric spheres. Each sphere has a specific net inward pressure $(C_r - R_r)$ on each unit area of its surface. The higher the value of $(C_r - R_r)$ is, the higher will be the curvature at that point. The total thrust on the sphere of radius r will be $4\pi r^2 (C_r - R_r)$ and this remains the same for all the spheres whatever the radius.

Identifying the integration constant C as the constant factor $(-4\pi r^2 (C_r - R_r))$ which remains the same for all the spheres, we have

$$e^{v} = \{ 1 - \alpha \, 4\pi r^2 (C_r - R_r)/r \} \dots\dots\dots\dots\dots\dots (14)$$

where α is a constant making e^{v} a number. Now ds^2 becomes,

$$ds^2 = \{ 1 - \alpha \, 4\pi r^2 (C_r - R_r)/r \} c^2 dt^2 -$$
$$\{ 1 - \alpha \, 4\pi r^2 (C_r - R_r)/r \}^{-1} dr^2 - r^2 (d\theta^2 + \sin^2\theta \, d\phi^2) \dots\dots\dots\dots (15)$$

Thus, we express the line element of the non-Euclidean space around the spherical body in terms of parameters associated with Space. This is a highly significant result that flows from our axioms for Space.

From eqn. (15), we can infer the following.

The space around a massive object becomes a gravitational pressure field whose strength $(C_r - R_r)$ goes on decreasing as we move away from the object. This gravitational pressure field at every point changes the nature of space at that point and makes the space non – Euclidean at that point. Thus, space gets distorted or curved by the presence of an object in it. This is the mechanism of space curvature. Since space and time are related and interconvertible, the gravitational pressure field at a point also distorts the flow of time at that point. Thus, we arrive at the concept of space – time curvature. g_{00} and g_{11} in equation (15) become a function of $(C_r - R_r)$. Once the space becomes non-Euclidean, the motion of an object in such a space is taken care of by the mechanics appropriate in such a curved space-time structure.

If $(C_s - R_s)$ is the net gravitational compressive pressure on the surface of the spherical body, $4\pi R^2 (C_s - R_s)$ will be the gravitational thrust on the spherical body. Since $(C_s - R_s)$ is the maximum value for the gravitational pressure, the curvature of space becomes maximum on the surface of the body. In other words, while a spherical body curves the space, the curvature at any point beyond the body is always less than the curvature at the body itself.

By our mass formula, we have for the mass m of the spherical body

$$m = \beta \, (4\pi R^2) \, (C_s - R_s) = \beta \, 4\pi r^2 \, (C_r - R_r) \dots\dots\dots\dots\dots\dots (16)$$

From eqns. (15) and (16), we get the relation.

$$ds^2 = (1 - \alpha \, m \, / \, \beta r) \, c^2 \, t^2 - (1 - \alpha \, m \, / \, \beta r)^{-1} \, dr^2 - r^2 \, (d\theta^2 + \sin^2 \theta \, d\phi^2) \dots (17)$$

To make e^v as a number, we take $\alpha = 2\beta \, G/c^2$, using which we get

$$ds^2 = (1 - 2 \, Gm/c^2 \, r) \, c^2 \, dt^2 - (1 - 2 \, Gm/c^2 \, r)^{-1} \, dr^2 - r^2$$
$$(d\theta^2 + \sin^2\theta \, d\phi^2) \dots\dots\dots\dots\dots\dots\dots\dots (18)$$

We've successfully arrived at the Schwarzschild metric relation. The key takeaways from our derivation are:

1. Our mass formula, not only yields Newton's law of gravity but also leads to the Schwarzschild metric in Einstein's theory – a testament to its remarkable versatility.

2. We have introduced and elucidated a potential mechanism underlying the space – time curvature.

3. Demonstrated that the curvature at any space – time point is always less than that causing distortion in space.

4. By initially expressing the line element in Space parameters and then incorporating our mass formula, we've been able to obtain the original expression of the Schwarzschild line element.

Summary

The Vethathirian theory of gravity introduces a groundbreaking model that enhances both Newton's and Einstein's theories of gravity. This model has profound implications for astrophysics, cosmology, and beyond. Built upon two fundamental axioms, it revolutionizes our understanding:

- It defines Space as the sole independent entity, devoid of external references. Time, Matter and Energy emanate from Space.

- Gravity is identified as the compressive attribute of Space, distinct from an inherent property of material particles.

- Dark energy is linked with waves in Space generated by the spinning motion of fundamental particles.

- The mass of an object is characterized as a measure of the net compressive force exerted by Space, on the object.

- This model proposes an underlying mechanism to comprehend the Space – Time curvature around a mass combined with our mass formula, the mechanism enabled us to derive the original Schwarzschild line element through a novel approach.

- It establishes, for the first time, a theoretical foundation demonstrating the equivalence between inertial and gravitational mass – a fundamental principle in physics and a cornerstone of Einstein's gravitational theory.

- It also derives, for the first time, Newton's law of gravity from basic axioms. Newton had originally deduced this law from Kepler's laws. Our derivation unveils fresh perspectives on the interplay of compressive and repulsive forces in the universe. It reveals that both the mass of an object and the gravitational constant (G) can be variables and time-dependent.

- By shifting the focus from particle-centric to Space-centric physics, this redefinition offers exciting prospects for delving deeper into the origins of cosmos. It has the potential to lead to groundbreaking discoveries that will reshape our understanding of fundamental physics, and consequently, our perception of Nature.

Vethathiri Maharishi

Vethathiri Maharishi (1911-2006) was an eminent Indian philosopher, spiritual leader, and yoga master known for his profound insights into the nature of the universe. One of his seminal contributions lies in formulating a unique perspective on fundamental concepts like Space, gravity, energy, and consciousness.

In the Vethathirian model of gravity, his philosophical ideas have been imbued with scientific rigor, supported by a robust mathematical framework. The model posits that the universe emerges from an infinite expanse of potential energy and consciousness, termed Primordial Space. Within this framework, gravity is understood not as a conventional force but as the compressive property inherent to Space itself.

Furthermore, Vethathiri Maharishi proposed that spinning particles give rise to a repulsive force, complementing the compressive influence of gravity. The interplay between repulsive and compressive forces is crucial in maintaining the equilibrium of the universe.

By merging Vethathiri Maharishi's philosophical insights with scientific methodology, the Vethathiri model of gravity provides a unique lens through which we can contemplate the fundamental forces that govern our universe. This model represents a tribute to his visionary thinking, offering a fresh perspective that challenges conventional paradigms in physics.

***

A Note on Cosmology

Einstein laid the foundation for cosmology with two fundamental principles.

1. The universe, on a large scale, exhibits homogeneity and isotropy. In simpler terms, it appears the same from any vantage point and in any direction. This led to the inference of a uniform matter density pervading the universe, consistently non-zero.

2. The spatial extent (radius) of the universe remains constant, independent of time. In this context, "spatial" refers to the entire space enclosed within the universe.

While both Newtonian and Einsteinian theories of gravity exclusively account for a single force gravity operating at cosmological distances, this poses a challenge. The attractive nature of gravity implies galaxies would inexorably gravitate toward each other, resulting in potential collisions and the eventual collapse of the universe. To resolve this predicament, Einstein introduced a second force acting at cosmic level into his equations termed the 'cosmological constant'. This adjustment was aimed to maintain a static (non-expanding) universe. At that time, Einstein held the belief that the universe was in a state of equilibrium, aligning with the principle of spatial independence from time mentioned above.

However, Edwin Hubble's empirical work conclusively demonstrated that the universe was expanding. This led Einstein to retract the cosmological constant, deeming it his 'greatest blunder'. Any theory in cosmology, which talks only of one force is bound to encounter similar problems and that also includes the Big Bang theory. It is in this context that in chapter III, we incorporate various components of the universe within a framework of 'Vethathirian concept of gravity' that includes the compressive thrust of Space and the repulsive thrust of Dark energy. A unique aspect of the Vethathirian concept of gravity lies in the in-depth exploration of the origin and functional role of dark energy. This distinctive feature leads us to regard this novel theory of gravity as a potential avenue for delving further into the realm of cosmological inquiry.

For our work in cosmology kindly refer to our book "Physics Redefined", Notion publishers, 2021

References

1. Alagar Ramanujam, G., Fitzcharles, K. and Muralidharan (2019) Indian Journal Of physics, 93, 959-963 *http://doi.org/10.1007/s12648-018-01364-9*

2. Alagar Ramanujam, G., Fitzcharles, K. and Muralidharan (2017) Journal of Modern Physics, 8,1067-1071 *https://doi.org/10.4236/jmp.2017.87068*

3. Arora, V., Taneja, R. and Alagar Ramanujam, G. (2019) International Journal of Trend in Research and Development, 6,472 *http://www.ijtrd.com/papers/IJTRD20708.pdf*

4. *Alagar Ramanujam,G., Padma Priya,D, Global Journal of Science Frontier Research: A Physics and Space Science, Volume 20, 10 https://globaljournals.org/GJSFR_Volume20/3-A-Theoretical-proof.pdf (2020)*

5. Vethathiri Maharishi, Unified Force, Vethathiri Publications, (Erode) 1996

6. Jammer, M., Concepts of Mass, Princeton Univ.Press, 2000

7. Wheeler, J.A. (1969) Relativity. Proceedings of the Relativity Conference in the Midwest, Cincinnati, 2-6 June 1969, 31-42. *https://www.springer.com/gp/book/9781468407235*

8. Andrew Janiak, (2004) Ed. Newton's Philosophical Writings: Cambride University Press.

9. Alagar Ramanujam, G & Vijay Arora: Physics Redefined (Notion Press) 2021.

10. Jayant Vishnu Narlikar: An Introduction to Cosmology (Cambridge University Press) 2002.

11. S.P. Puri: General Theory of Relativity, Dorling Kindersley (India) Pvt. Ltd.(2013)

12. Albert Einstein: Relativity (General Press) 1916.

Physics Behind the Dark Matter, Dark Energy and the Inflationary Expansion of the Universe

G Alagar Ramanujam[1]*, K Fitzcharles[2] and S Muralidharan[3]

[1]N.G.M. College, Pollachi, India

[2]Vethathiri International Academy, Chennai, India

[3]Srinivasa Ramanujan Centre, SASTRA Deemed to be University, Kumbakonam, India Received: 31 May 2018 / Accepted: 28 September 2018

Abstract: In recent papers, we advanced a set of axioms for Space and using these axioms, the nature of Dark Energy was discussed. In this paper, we show that the inflationary nature of expansion of the universe emerges from our axioms. We further throw light on the question of whether it is necessary to introduce the concept of Dark Matter to explain certain anomalies regarding the motion of certain galaxies.

Keywords: Dark Energy; Dark Matter; Cosmology; Inflationary expansion; Acceleration

PACS Nos.: 98.80.-k; 95.35.?d; 98.80.-k; 04.50.kd; 98.80.Bp

*Corresponding author, E-mail: gravity2003@gmail.com

1. Introduction

For the past two decades, more and more fresh and chal- lenging problems have been emerging in physics. The study of the evolution of the universe is posing nowadays many challenging problems. It is very difficult to resolve them using the existing concepts in physics. In this context, we see in recent times fresh proposals like Dark Matter, Dark Energy, Higgs boson, self-evolving universe, being advanced. It is now clear that to develop a satisfactory theory for the evolution of the universe, we need fresh insights into the way nature works.

The reality before us is very simple: *Mass subsists within Space.* In other words, a massive universe with a finite radius is floating in an all-pervading, boundary-less Space. In science, space generally means space interval between two objects, but not the primordial Space in which universe is floating. Every particle or a system of particles of the universe is massive, and hence a force of sufficient strength has to act on it in order to move it. We are still at a loss to understand how that particle acquires its massive- ness or simply what is mass. Without a basic and logical understanding of what mass is, our understanding of the universe will remain incomplete.

While Newton discussed what mass does (Mass always opposes the force applied on it: $a = F/m$), he never defined what mass is. That he considered the mass of a particle as a constant is a severe limitation of his theory.

While Einstein brought out the equivalence between mass and energy ($E = mc^2$), he also never defined what mass is. He improved upon Newton's concept by consid- ering that the mass of a particle is a variable with respect to its velocity. When he came to the general theory of rela- tivity, he discussed the impact of mass on Space.

Since there is no clear understanding of what mass is, we are at a loss to fully understand what the universe is and how it evolves. To have a proper understanding of the evolution of the universe, it is necessary for us to under- stand the relation between a particle and Space.

So, before venturing to give a theory for the universe and its evolution, it is imperative on our part to know

(a) What is Space?

(b) What is Mass?

(c) What is the relationship between Space and Mass?

In the year 2014, we proposed a new theory to address above questions wherein we considered Space (Primordial Space) as the absolute, fundamental state and we defined a particle as a collection of infinitesimal quantum excitations in Space. Any theory has to begin from a set of axioms. We give below our axioms framed by us in the year 2014 for Space, which directly bear on the above questions.

1.1. Axioms

We give below our axioms about Space and discuss briefly the implicit consequences derived therefrom [1–3].

1. Space is all-pervading and is endowed with potential energy. It has the property of constant self-compres- sion and continually exerting compressive pressure on every system in it.

2. Self-compression results in the formation of infinites- imal spinning quanta of Space, called "formative dust". Due to the surrounding compression, dust is forced into formation of discrete groups we know as fundamental particles; every group of dust formed by the surrounding compressive pressure of Space has a spin and hence becomes a source of a radial field with a repulsive force at every space–time point.

The first axiom defines the attributes of Space, especially its property of acting on itself. This self-reflexive property of Space has become a hot topic in recent works in Astrophysics [4–6]. The second axiom defines the relationship between Space and a particle. An infinitesimal, imperceptible, spinning quantum of Space is here called a formative dust. An association formed of innumerable dust is called a fundamental particle.

A significant consequence of the above axioms is that they lead us to an understanding of what mass is. The surrounding Space exerts a compressive thrust on every system of particles, and this compression is partially countered by the repulsive pressure due to the spin of the particles in the system. There is thus a net compressive thrust on every system, or in other words, every system is in the grip of Space. Due to this grip, one finds it difficult to move it; then one says, "The system is massive or it has a mass". If 'A' is the area of a system, then the mass 'm' of the system is given as [1]

$$m = \beta A (C - R) \tag{1}$$

where C is the compressive pressure on the system due to Space and is the cause of Gravity. R is the repulsive pressure on unit area of the system due to the spin of the particles comprising the system and β is a universal constant.

Using the above relation for the mass of a system, we gave the first ever derivation for the Newton's law of gravity from basic axioms. We point out that Newton did not derive his law of gravity from axioms, but only deduced it from Kepler's laws of planetary motion. We give below the salient features of our derivation of the law of gravity.

If m_1 and m_2 are the masses of two particles with a distance r between them, $m_1 = \beta A_1(C_1 - R_1)$ and $m_2 = \beta A_2(C_2 - R_2)$ where A_1 and A_2 are the surface area of the masses.

Due to the combined effect of the compressive force exerted on the particles by the Space surrounding them and the repulsive force between them due to their spins, the force experienced by m_1 towards m_2 was shown to be

$$F_1 = k_2 m_2 = 4\beta \Pi r^2 \tag{2}$$

and the force on m_2 towards m_1 was shown to be

$$F_2 = k_1 m_1 = 4\beta \Pi r^2 \tag{3}$$

where k_1 and k_2 depend on the size of the particles and their relative orientation with respect to the other particle. Treating F_1 and F_2 as action and reaction, we have $F_1 = F_2$ and hence

$$k_2 m_2 = k_1 m_1 \tag{4}$$

Solving the above, we have $k_1 = \mu(r)m_2$ and $k_2 = \mu(r)m_1$

Therefore $F_1 = F_2 = F = \mu(r)\, m_2 m_1 / 4\beta \Pi r^2$

$$= G(r)m_1 m_2 = r^2 \tag{5}$$

where $G(r) = \mu(r)/4\pi\beta$.

In a dimension of our solar system, G is taken as a constant. But across the galaxies, G is certainly a variable.

By choosing a proper function for $G(r)$, we can explain the anomalies observed regarding the velocities of galaxies in the clusters without introducing the concept of Dark Matter. It is gratifying to note that such an attempt has been made by Stabnikov and Babailov [7] in their recent paper wherein they have taken $G(r) = (G + \delta r)$ and have shown that the anomalies associated with the motion of stars and galaxies can be explained without introducing the notion of the Dark Matter. (G is the usual gravitational constant and δ is $2{:}7 \times 10^{-31}$ m^2 kg^{-1} s^{-2}). The variation of G has also been indicated by several authors through various consid- erations [8–10]. By choosing a suitable expression for $G(r)$, the concept of Dark Matter can be dispensed with.

2. Dark Energy and the accelerating expansion

Recent observations in Astrophysics and Cosmology clearly indicate that our universe is expanding with an acceleration [2, 3, 11–15]. Because of the Newtonian mutual attraction existing between any two objects of the universe, one would expect the universe to contract more and more. But the observations of Hubble [16] clearly established that the universe is expanding. This discovery of the expansion of the universe led in recent years to the conjecture that there must exist a kind of energy pushing galaxies away from each other despite the Newtonian attraction between them. Since the source of such energy is still unknown, it is called Dark Energy. By using our basic axioms, we can explain the source of Dark Energy responsible for the expansion of the universe.

2.1. Theory of Dark Energy

One can consider the matter contained in the universe as a perfect cosmic fluid of particles whose density remains the same at every space–time point [8]. The Space permeates through the universe. By our axioms, more and more par-ticles are produced continuously of Space and by Space and to accommodate these particles the boundary of the universe has to expand. Further, every particle of the fluid continuously loses energy to Space and thereby the spin of the particle gradually decreases over time. The energy gained by Space results in a wave front in the space–time fabric diverging from the particle. These waves in space– time exert an outward pressure on the boundary of the universe. The energy associated with such waves in space– time is what we call Dark Energy. Since we now know the source of the Dark Energy, it is no longer dark for us and we call that energy 'Repulsive Energy' operating at every space–time point. Taking the above factors into account, we derived two equations which resemble the famous Friedmann cosmological equations. One of them is:

$$\ddot{r} = -\left(\frac{4\pi G}{3}\right)r\rho_{M} + \frac{(\rho_{M} + \rho_{E})c^{2}}{3} \tag{6}$$

$$H^{2} = \frac{(\rho_{M} + \rho_{E})c^{2}}{3r} - \frac{4\pi G}{3}\rho_{M} \tag{7}$$

ρ_M and ρ_E are matter density and Dark Energy Density of the universe. $r\epsilon$ is the force on the unit mass placed at the edge of the universe whose radius is r and H is the Hubble factor.

It is relevant here to quote the work of Murad Shibli wherein he states: "The negative pressure of the Dark Energy is equal to the Energy Density. It is found that Dark Energy is a property of the space–time itself. Based on the fluid nature of Dark Energy, the fourth law of thermody- namics is presented" [15].

We discuss below the variation of the acceleration over the radius of the universe. For this we make use of our cosmological Eqs. (6, 7) derived from our axioms. Con- sidering one of the above two equations, we have

$$\ddot{r} = -\left(\frac{4\pi G}{3}\right) r\rho_M + \frac{(\rho_M + \rho_E)c^2}{3}$$

The acceleration becomes zero at r_e where

$$r_e = \frac{(\rho_M + \rho_E)c^2}{4\pi G\rho_M} \tag{8}$$

Beyond r_e, by Eq. (6) the compressive force exerted by Space on the universe exceeds the repulsive force and hence the velocity starts decreasing and finally becomes zero. Then the contraction of the universe starts.

Solving Eq. (6),

$$r = r_e - (r_e - R)\cos \alpha t \tag{9}$$

where r_e is the radius at which the acceleration becomes zero and R is the finite minimum value of r at $t = 0$ and $\alpha^2 = 4\pi G\rho_M/3$ (Here $t = 0$ does not mean the zero in absolute time, but the time frame of our observed presently expanding universe). The universe with the radius R needs a special mention. A stable group of formative dust is what we call fundamental particle in science. With the formation of the fundamental particle, the universe begins. A group of fundamental particles compressed by the surrounding compressive pressure of Space forms a system. With more and more fundamental particles pushed into it, the system gradually assumes a spherical shape with perfect homo- geneity. The density of the fundamental particles in the sphere is the same everywhere. Now the radius of the sphere is indicated by R. At this stage, the repulsive wave flux inside the sphere dominates the compressive thrust on the sphere by the Space and hence the sphere is in expansion mode that we now observe. The early universe with radius R is what the standard Big Bang Model calls the primordial egg which exploded billions and billions of years ago. While the standard big bang model does not existence, we have a mechanism to understand how that primordial egg came into existence.

From Eq. (9) we can get

$$V = \alpha \left\{ \left(k^2 - r_e^2(1-q)^2 \right) \right\}^{\frac{1}{2}}$$

$$\frac{dv}{dr} = \frac{\alpha r_{e(1-q)}}{\left\{ \left(k^2 - r_e^2(1-q)^2 \right) \right\}^{\frac{1}{2}}} \tag{10}$$

where $k = (r_e - R)$ and $\rho = r/r_e$. Equation (10) gives the variation of the expansion velocity over 'r'. The equation indicates that the early universe of very low radius had an inflationary rate of expansion over r and that rate is steadily decreasing as r increases as shown in Fig. 1. It is signifi- cant that the inflationary mode of expansion of the early universe [17] emerges from our theory as a consequence of our axioms about Space. It must be mentioned here the observations of Hawking [6] regarding the inflationary expansion of the universe: "The idea that such an episode of inflation might have occurred was first proposed in 1980, based on considerations that go beyond Einstein's theory of general relativity and take into account aspects of quantum theory. Since we don't have a complete quantum theory of gravity, the details are still being worked out, and physi- cists aren't sure exactly how inflation happened". As already explained, now we have a mechanism to under- stand how the inflation happened.

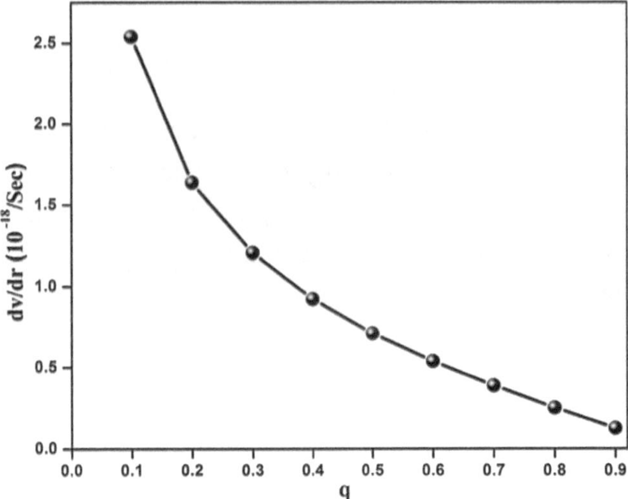

Fig. 1: dv/dr decreases as q increases

In drawing the above graph, we have taken $(r_e - R) \approx r_e$. The graph clearly indicates the inflationary nature of expansion. The above graph was drawn by assigning values for ρ_M and ρ_E given for them in Ref. [15]. $\rho_M = 0.54207 \times 10^{-26}$ kg/m^3 and $\rho_E = 1.2622 \times 10^{-26}$ kg/ m^3.

According to Eq. (8) of the paper,

$$r_e = \frac{(\rho_M + \rho_E)c^2}{4\pi G \rho_M}$$

where r_e is the distance at which (dv/dr) becomes zero. With values 0.54207×10^{-26} kg/m^3 and 1.2622×10^{-26} kg/m^3, respectively, for ρ_M (Matter Den- sity of the Universe) and ρ_E (Dark Energy Density of the universe) taken from Ref. [15], the value of r_e turns out to be 3.55×10^{26} m (or 37.53 billion light years). This value for r_e will get a refinement when we take quantum mechanical and relativistic effects into our theory.

At $r = r_e$, (dv/dr) becomes zero and thereafter the compressive power of gravity becomes greater than the repulsive force due to Dark Energy. Consequently then, the expansion of the universe will be opposed and the radius of the universe will finally get back to R.

In Fig. 1 (dv/dr) is steadily decreasing as ρ increases. On the x axis ρ (r/r_e) is unitless, and on the y axis the unit is 10^{-18}/s.

The above graph agrees fairly well with the graph obtained by Mishra and Vadrevu [18] for the variation of the Hubble factor with time as given in Fig. 2.

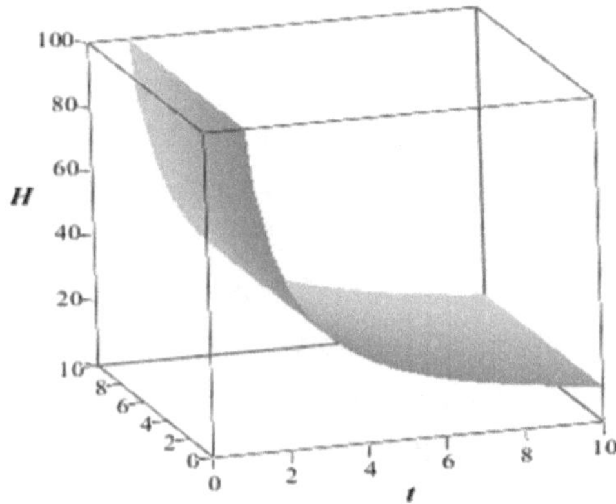

Fig. 2: *H* is steadily decreasing as *t* increases

Both our graph and the graph obtained by Mishra and Vadrevu clearly indicate that as r or t increases, the velocity v increases, but the rate at which *v* increases slowly decreases.

3. Conclusion

From our axioms given for Space, we first derived a for- mula for Mass and using the mass formula, Newton's law of gravity [1] was derived. The mass formula given in Eq. (1) is highly significant. In that equation, the right-hand side represents the interaction between the particle through its spin and the compressive nature of Space. This inter- action brings to mind the mechanism proposed by Peter Higgs [19] to explain how particles acquire their mass through the field introduced by him.

Our axioms gave us Friedmann like cosmological equations by using which we obtained a modified Hubble's law [8]. In this paper, we use our axioms to explain the source and the repulsive nature of Dark Energy and obtained the inflationary expansion of the early universe. All our results can further be improved by introducing relativistic and quantum principles into our work.

In addition, our future work will introduce the decay of energy particles back into formative dust as a further out- come of our axioms. This may well lead to more refine- ment of the model presented in this paper. More in this direction will be published elsewhere. Since our axioms are based on the philosophy of a great thinker of Tamil Nadu, India, Shri. Vethathiri Maharishi, we are collectively labelling all our results arising out of our axioms as part of Vethathirian Cosmology.

The two axioms framed by Einstein for light brought out many results which could not have come from Newtonian physics. The relation $E = mc^2$ is beyond the scope of Newtonian physics but arises naturally as a consequence of the axioms about light. We have a similar situation here. The present paper is a consequence of our axioms about Space, which is much subtler than light. The inflationary expansion of the early universe, as Stephen Hawking [6] remarked, is beyond the scope of General relativity. According to him: "The idea that such an episode of inflation might have occurred was first proposed in 1980, based on considerations that go beyond Einstein's theory of general relativity". It is highly gratifying to note that the inflationary expansion of the early universe emerges from our axioms.

Acknowledgements The authors place their deepest reverence to Shri. Vethathiri Maharishi for having taught us his philosophy of Space in all details and in all depth. They remain highly grateful to Mrs. and Mr. Raj Taneja of Chicago, Mrs. and Mr. Vijay Arora, of New Jersey, Mrs. and Dr. Panneer Selvam of Florida International University, Miami, Mrs. A. Thayar of Pollachi, Mrs. and Mr. Vikram Kannan of New Jersey, Mrs. and Mr. Babu Prasath of Pollachi, Mrs. M. Vimala, Miss. M. Devatharshini, Mr. M. Srivatssan, and Mr. M. Aswin of Kumbakonam for their constant interest in this work.

References

[1]   G Alagar Ramanujam and K Fitzcharles *Int. J. Trends Phys. Sci.* 3 76 (2014)

[2]   G Alagar Ramanujam and K Fitzcharles *Int. J. Trends Res. Dev.* 3 215 (2016)

[3]   G Alagar Ramanujam, K Fitzcharles and S Muralidharan *J. Modern Phys.* 8 1067 (2017)

[4]   J P Merrison *Int. J. Astron. Astrophys.* 6 312 (2016)

[5]   S G Turyshev *Physics-Uspekhi* 52 1 (2009)

[6]   S Hawking and I Mlodinow *The Grand Design* (London: Ban- tam press) (2010)

[7]   P A Stabnikov and S P Babailov *J. Astophys. Aerospace Tech- nol.* 5 1 (2017)

[8]   J P Merrison *Int.J. Astron. Astrophys.* 6 312 (2016)

[9]   N A Khomane and G V Kolhe *J. Appl. Phys.* 9, 15 (2017)

[10] A E S Abou Layla *J High Energy Phys.* 3 248 (2017)

[11] A K Yadav *Astrophys. Space Sci.* 361 276 (2016)

[12] Z G Sha and R Xiu *J. Astrophys. Aerospace Technol.* 5 1 (2017)

[13] A G Riess et al. *Astrono. J.* 116 1009 (1998)

[14] M S Turner *The third Astro. Sympo.* 666 1 (1999)

[15] M Shibli *IEEE* 1-4244-1057-6/07/2007

[16] E Hubble *PNAS* 15 168 (1929)

[17] Guth H Alan *The Inflationary Universe (Perseus Books, USA)* p 233 (1997)

[18] B Mishra and S Vadrevu *Astrophys Space Sci.* 26 1 (2017)

[19] P W Higgs *Phys. Lett.* 12 132 (1964)